DEVELOPMENT OF BOTANY
IN
SELECTED REGIONS
OF
NORTH AMERICA BEFORE 1900

DEVELOPMENT OF BOTANY
IN
SELECTED REGIONS
OF
NORTH AMERICA BEFORE 1900

Edited with an Introduction by
Ronald L. Stuckey

NYT

ARNO PRESS
A New York Times Company
New York • 1978

Editorial Supervision: MARIE STARECK

———◆———

Reprint Edition 1978 by Arno Press Inc.

Arrangement and Compilation Copyright © 1978
   by Arno Press Inc.

Introduction Copyright © 1978 by Ronald L. Stuckey

The "Introduction" by John W. Harshberger, reprinted
from The Botanists of Philadelphia and Their Work,
was reprinted from a copy in The University of
Illinois Library.

BIOLOGISTS AND THEIR WORLD
ISBN for complete set: 0-405-10641-6
See last pages of this volume for titles.

Manufactured in the United States of America

Publisher's Note: The selections in this anthology
were reprinted from the best available copies.

———◆———

Library of Congress Cataloging in Publication Data

Main entry under title:

Development of botany in selected regions of
   North America before 1900.

   (Biologists and their world)
   Reprint of various articles published in
journals from 1879-1909.
   1. Botany--United States--History--Addresses,
essays, lectures. 2. Botanists--United States--
Biography--Addresses, essays, lectures.
I. Stuckey, Ronald L. II. Series.
QK21.U5D48   581'.973   77-81126
ISBN 0-405-10722-6

INTRODUCTION

The development of botany in North America has many facets. It involves the personalities of individuals, their travels and explorations, the acquisition and fate of the plants they obtained, the establishment of herbaria and botanical gardens, the organization of botanical societies and museums, the teaching of botany in schools and colleges, and the publication of observations and researches. All of these facets have been instrumental in the perpetuation of the knowledge of plants as it passed from teacher to student, from scientific paper to textbook, from professional to amateur, and from colleague to friend. In this manner future generations learn from the past and carry the science forward.

Many of the above mentioned subdisciplines are sampled and discussed in the papers that have been selected here for reprinting. They represent papers that were written about 1900 on the development of botany in various regions of the United States, principally in the eastern United States, where most of the pioneer botanical work had been accomplished to that time. The regions chosen were representative centers of botanical activity, and perhaps it was merely coincidental that each of these areas had a botanical history written of it at that time. Examination of these individual regional histories shows them to vary in the quality and quantity of information presented. The reader should bear in mind that viewpoint while studying their contents. However, in all instances[1] the authors of the essays use individual botanists as their central focus and discuss their major accomplishments. These biographical accounts, in essence, reveal the botanical development. Some individuals worked in and were influential in more than one region and so a discussion and evaluation of their work appears more than once. The paper by Frederick Brendel on the "Historical Sketch of the Science of Botany in North America from 1635 to 1840," although written about 20 years before 1900, has been selected as the introductory article because of its broad overview of the history of descriptive botany. In that sense Brendel's paper serves as a good foundation for the papers that trace the botanical history for each of the several regions. These papers were authored by competent local botanists who knew first-hand the history of the botany in their region.

The development of American botany from its beginning in the seventeenth century to the opening of the twentieth century was a period of considerable advancement in the naming, cataloguing, describing, and classifying of the plants for this continent. Botany was not advanced by professional people, but rather by dedicated individuals who devoted their leisure time to the study of plants. Among the earliest botanist-naturalists were travelers from Europe who were sent to America to discover its economic plants and to enrich the gardens of Europe. They sold living plants, seeds, and dried specimens to aid in defraying their expenses. In the course of searching for these "useful" plants, all plants, including the "non-useful" were examined, collected, and identified. During this process many excellent taxonomic works that classified the plants resulted. Although this scientific botany was of secondary interest at the time, scientific botany grew out of these explorations. The plant collections that resulted laid the foundations for the great American herbaria, documented the published local and regional floras, and paved the way for the study of systematic botany in North America during the twentieth century. Many of our great naturalists are identified with this pioneer period of plant exploration in the vast frontier. Among those from Europe were André Michaux, the French explorer, Frederick Pursh, a German immigrant and gardener, Thomas Nuttall, the English printer and naturalist, and many other traveling naturalists, François André Michaux, C. C. Robin, David Douglas, John Bradbury, and Thomas Drummond, to name a few. The American Quaker farmer-botanist John Bartram did much to make American plants known in Europe through his exchanges of seeds, plants, letters, and conversations with his friends abroad. Later, Amos Eaton, who has been referred to as "the greatest popularizer of natural science that America has ever known," became well-known for his teaching of botany and his practical botanical manuals that went through eight editions.

The earliest teachers and extensive writers on botany, however, were usually physicians because they had received scientific training in the medical schools. The study of plants was essential to the understanding of *materia medica*. In this category, outstanding medical teachers of botany and *materia medica* that deserve mention are Benjamin Smith Barton, William P. C. Barton, Lewis Caleb Beck, Jacob Bigelow, Asa Gray, David Hosack, Samuel L. Mitchell, John L. Riddell, Charles Wilkins Short, John Torrey, and Benjamin Waterhouse. During the nineteenth century, a shift occured from the part-time teaching of botany, as performed by Benjamin Smith Barton, to a full-time professorship in botany that was held by Asa Gray which served to mark the end of this era. Other physicians not associated with medical schools as teachers, but who worked diligently in botany along with their extensive medical practice were such individuals as William Baldwin, Alvin Wentworth Chapman, William Darlington, George Engelmann, Edwin James, Charles Mohr, Charles Pickering, and Adolph Wislizenus.

Clergymen were also involved in the development of American botany, men such as Moses A. Curtis, Manasseh Cutler, Chester Dewey, Auguste B. Langlois, Henry Muhlenberg, John Lewis Russell, and David Lewis von Schweinitz. They apparently had ample time to study plants in the countryside as they made visits to members of their congregations.

The beginning of American botany is a story of the progress of the written flora. Local floras, such as those by Manasseh Cutler for New England, Henry Muhlenberg for Lancaster, Pennsylvania, John Torrey for New York, William P. C. Barton for Philadelphia, Stephen Elliott for the Carolinas and Georgia, John L. Riddell for the states immediately west of the Allegheny Mountains, and Charles W. Short for Kentucky, are representative examples. Floras of a larger or continental scale were first prepared by André Michaux and Frederick Pursh and published in Europe, but later floras of continental scope were published in America, such as those by Thomas Nuttall, John Torrey and Asa Gray, and Constantine Samuel Rafinesque.

The development of herbaria were likewise important—first as a means of identifying the living plants, then as private collections, and then as an institutional enterprise. Among the earliest herbaria were those of the American Philosophical Society and the Academy of Natural Sciences, both in Philadelphia, the Charleston Museum in South Carolina, and the William Darlington Herbarium in West Chester, Pennsylvania. Later John Torrey's herbarium at Columbia College, Asa Gray's herbarium, and that belonging to George Engelmann, each became the nucleus of the great herbaria that have since been developed at the New York Botanical Garden, the Gray Herbarium of Harvard University, and the Missouri Botanical Garden, respectively.

Nurseries and botanical gardens have played a significant role in the advancement of botany in the United States. Important early nurseries were developed by William Prince of Long Island and by John Lyon, Bernard McMahon, and Thomas Meehan of Philadelphia. Among botanical gardens, the early ones were those established by John Bartram in Philadelphia, Humphrey Marshall near Philadelphia, André Michaux in New Jersey and Charleston, South Carolina, and David Hosack in New York City. Later, the Harvard Botanical Garden, founded by Asa Gray, Henry Shaw's Garden, which became the Missouri Botanical Garden, and the New York Botanical Garden, founded by Nathaniel Lord Britton, served the botanical community as leading institutions in many phases of botanical science well into the twentieth century.

Societies for the exchange and promotion of ideas, such as the American Philosophical Society and the Academy of Natural Sciences, both of Philadelphia, were both at a very early date instrumental in bringing naturalists and potential nalturalists together. The Botanical Society of Pennsylvania and the Philadelphia Botanical Club later became strong organizations in the Philadelphia area. In New England were organized the New England Society for Promoting Natural History, the Boston Society of Natural History, the Natural History Survey of Massachusetts, the American Academy of Arts and Sciences, and the Essex County Natural History Society. In New York, the Lyceum of Natural History and the Torrey Botanical Club came into existence. The Washington Botanical Society was founded in the District of Columbia. In the South, the Charleston Library Society was an important scientific organization. These early natural history and botanical societies did much to pave the way for scientific botany and diffuse botanical knowledge through their meetings and publications.

After the early portion of the nineteenth century had passed, American botanical works, which had been published in Europe, were increasingly being published in this country. This new procedure provided more opportunities for individuals to publish the results of their efforts and to preserve a record of their observations and studies. When the study of botany in America reached this stage, it was ready to stand on its own merits. Truly, the pioneer period of American botany was a period of exploration, of foundation laying, and one of broadening horizons into many areas of advancement. This series of papers sketches these developments and horizons and should prove useful to both botanists and non-botanists who desire an overview of the progress of American botany from its inception to the twentieth century.

Ronald L. Stuckey
Associate Professor of Botany
The Ohio State University
Columbus, Ohio 43210
October, 1976

---

[1]In *The Botanists of Philadelphia and Their Work* by John W. Harshberger, Philadelphia, 1899, only the 40-page introduction which gives a general overview of botanical developments in that city is reproduced here. The main portion of the book comprising 132 biographical accounts of botanists is not reprinted.

CONTENTS

Cocks, R. S.
Historical Sketch of the Botany of Louisiana (Reprinted from
Proceedings of the Louisiana Society of Naturalists, 1897-1899),
New Orleans, 1900

Spaulding, Perley
A Biographical History of Botany at St. Louis, Missouri, [Parts]
I-IV (Reprinted from *The Popular Science Monthly*, Vols.
LXXIII and LXXIV), New York, 1908 and 1909

PART I

OVERVIEW

HISTORICAL SKETCH OF
THE SCIENCE OF BOTANY
IN NORTH AMERICA FROM 1635 TO 1840

Frederick Brendel

HISTORICAL SKETCH OF THE SCIENCE OF BOTANY IN NORTH AMERICA FROM 1635 TO 1840.

BY FREDERICK BRENDEL.

A HISTORY of the science of botany in North America means not in this sketch a history of that science in all its branches, but rather the history of traveling and local collectors, and of descriptive botany so far as it concerns American plants. For until Prof. A. Gray's popular book, "How Plants Grow" appeared in 1858, not a single work of any importance was published in this country, either on anatomy or on the physiology of plants, not even a single one of the many systems ever proposed had its origin in America. And yet the labors of American and foreign scientists in America contributed their large share to the advancement of science. They furnished the material for the work in all the other branches of botany, and particularly in the geography of plants. Most of them did a toilsome work, exposed in the wilderness to manifold fatigues and perils; many died far from home on the glorious battlefield of science, as it were, sword in hand; some a violent death, others swept away by a pernicious climate.

1635-1800.—It was in 1635 that the first book on North American plants ever written, was published by Jacques Philippe Cornut, a French physician. He described Canadian plants brought over to Europe, in a book entitled: *Canadensium Plantarum Historia.* It is illustrated by good drawings, most of the species being recognizable at first sight, though the names given are quite different from those now in use. But the work does not contain, as might be inferred from the title, Canadian plants only, but also some others from Spain and the Orient. Not until thirty-seven years afterward, in 1672, was another account of American plants given by John Josselyn, in a book entitled *Rariora Novæ Angliæ*, and in 1674, in an account of two voyages in New England.

At the same time, in 1672, Wm. Hughes published in London, The American Physician, or a Treatise of the Roots, Plants, etc.

In Ray's Historia Plantarum, 1688, second volume, we find a "Catalogus plantarum in Virginia observatarum," by John Banister, an English missionary and botanist, who came, in 1680, to Virginia, where he made his collections. The same catalogue was republished, in 1707, in Petiver's Memoirs for the Curious. About the same time an Englishman, Wm. Vernon, and a German, David Krieg, collected, in Maryland, several hundred new species, which they sent to Ray, Petiver, Sir Hans Sloane and others.

James Petiver, a London apothecary, described, in 1706, in " Pterigraphia Americana," some North American ferns, and Leonard Pluckenet, a London physician who lived from 1642 to 1706, figured many North American plants in " Almagestum Botanicum," 1696, and " Almatheum Botanicum," 1705.

The same year came the English naturalist, John Clayton, (1685-1773), to Virginia, where he made his collections, afterwards described by Gronovius, a distinguished Dutch botanist at the University of Leyden, in Holland. His " Flora Virginica Exhibens Plantas, quas J. Clayton in Virginia collegit," was published in 1743, and a second edition by Gronovius, the son, 1762, augmented by observations of Clayton, Colden, Mitchell and Kalm.

From 1712 to 1719, Mark Catesby, another English naturalist, collected in Virginia. A second time he started from England and arrived, in 1722, in South Carolina. He traveled three years in that State, in Georgia and Florida, visited the Bahamas and came back to England in 1726, where he published from 1731 to 1743, the valuable work, " The Natural History of Carolina, Florida and the Bahama islands," two volumes in folio and a supplement with two hundred and twenty colored plates. The descriptions are in English and French ; a German edition was published in 1750. After his death (1749) was published his Hortus Britano-Americanus, in which he described the trees and shrubs of the British colonies in North America adapted to the soil and climate of England. London, 1763.

Here may be mentioned a natural history of North Carolina, by Brickell, in Dublin, 1737.

The Swedish naturalist, Peter Kalm, explored the eastern part of Pennsylvania, New Jersey, New York and Canada during the years 1748 to 1751. He was sent by the Swedish government at the proposal of Linnæus, whose pupil he was. The original motive was the American mulberry (*Morus rubra*), which was known to grow as far north as Canada, in a climate similar to that of Sweden. It was intended to acclimatize the tree and to introduce the culture of the silkworm in Sweden. The extensive botanical collections were worked up by Linnæus and embodied in his herbarium, where they exist yet.

Linnæus published, 1753, the first edition of his "Species Plantarum," in which he described 5938 species, all that was known at that time, and of which 5323 were phænogamous. In all his writings the number of species he was acquainted with was 8551 (7728 phænogamous and 825 cryptogamous). Amongst these are 1075 species either common to the eastern and western continents or only North American, inclusive of the Arctic regions and exclusive of Mexico.

Another contributor to the herbarium of Linnæus was Cadwallader Colden, Lt. Governor of New York, who was born in Scotland, 1688, and died in New York, 1776. The catalogue of his collection is published in "Acta Societatis Scientiarum Upsalensis, 1743–1744. His daughter, Miss Jenny Colden, wrote a Flora of New York, with drawings, the manuscript of which she transferred, before her death, to Wangenheim; afterwards it was incorporated into the Banksian Library, but never published.

Some Canadian plants were described by the French missionary, Xavier de Charlevoix, in his " Histoire et description générale de la nouvelle France, 1744." About the same time, John Bartram (born 1701, died 1777 in Philadelphia) traveled to Lake Ontario, and published his valuable observations in 1751. His son, Wm. Bartram (1739–1823), was the first to extend his botanical excursions to the Southern Alleghanies. He left Philadelphia in 1773, traveled through Florida and Georgia to the Cherokee country, and went through Alabama to Mobile in 1776. He published his voyage in Philadelphia, 1791. This valuable work was afterwards translated into the German, in 1794, and in French, in 1799.

John Mitchell, an American physician in Virginia, published additions to Linnæus' first edition of Genera Plantarum in his " Dissertatio de Principiis Botanicorum."

Reinhold Forster, naturalist to Cook's second expedition, compiled a catalogue of the plants of North America, in 1771, without descriptions.

A number of species, common to the Southern States and the West Indies, were made known by the important discoveries of Chas. Plumier (1690–1695), of Nic. Jos. de Jacquin (1754–1759) and of Olaf Swartz (1785–1789). Only the latter spent a year on the North American continent before he went to the West Indies, where he alone discovered and described 850 new species. The first came from France, the second from Germany, the last from Sweden.

At that time Humphrey Marshall made the woody plants his special study. He published his " Arboretum Americanum," containing 276 species, in Philadelphia, 1785, which in 1788 was republished in Germany. The German foresters took a special interest in the matter, as many American woody plants had already found their way into the German nurseries, and by some experiments made it was known that for many purposes some sorts of American timber were superior to the indigenous, and as the greater demand for fuel in some industrial districts resulted in a final scarcity of wood, they thought that the cultivation of American timber in the German forests would be, by its more rapid growth, of great advantage.

Captain[1] Wangenheim, of the Hessian troops, afterwards Prussian forest-officer, studied, during his eight years service in America, the timber of this country with regard to its usefulness and practicability of culture. In 1781 he published descriptions of some North American trees, and after his return to Germany a larger work with drawings, 1787. There are many good observations on the soil and climate and their influence upon the culture of the different species, but the drawings are inferior, and in one there is a great error which is worthy of notice. On plate 18 is figured the leaf of *Carya olivæformis* and what is intended to represent its fruit, but looks rather like a pea nut. Probably he had never seen the nut, which he described as " kidney shaped," though he examined, in Wm. Prince's nursery at Flushing, on Long Island, the young tree not yet bearing. As he was eager to obtain the fruit, somebody by mistake, or perhaps for jest, may have given to him a pea nut for a pecan nut, which he drew. He gives a short history of the tree, which was unknown in the English colonies until the peace of 1762, when by chance some fur-traders brought a small number of the nuts to New York. Wm. Prince planted (1772) thirty nuts and raised ten plants, which (except two retained for propagation) he sold to England at ten guineas a piece.

[1] Here may be corrected an error in the preface of Torrey's Flora of New York. Wangenheim was not a surgeon nor an M. D.

In 1783, was sent over to America from Vienna, a scientific expedition under the charge of Prof. Marter, assisted by Dr. Stupicz, two gardeners and one designer. From Philadelphia they made excursions in Pennsylvania, to Virginia and Carolina. In the latter State, Marter met Dr. Schoepf, surgeon in the service of the Margrave of Ansbach, another German petit-tyrant who sold his poor subjects to the English. Both made together an excursion to Florida and the Bahamas. Marter brought large collections to Vienna, and Dr. Schoepf afterwards published a "Materia Medica Americana," Erlangen, 1787, and his "Travels Through the North American States," Erlangen, 1788.

An Italian nobleman, Luigi Castiglioni, traveled from 1785 to 1787 in the Eastern States, and published, 1790, his "Viaggio negli Stati Uniti del America Settentrionale," in two volumes. The latter half of the second volume contains observations on the useful plants. Like Wangenheim he gives to his countrymen some hints in regard to acclimatation. He describes most of the eastern and southern woody plants and gives a few good drawings (*Franklinia alatamaha* of Marshall, now *Gordonia pubescens*, *Quercus banisteri* and *Rhus venenata*). He made himself acquainted with the scientific men of the country, and in a passage (p. 163 second volume) where he objects to the assertion of Raynal (Histoire Philosophique et Politique), that America has never produced a single prominent man, be it in science, art or any other branch, he names, after mentioning a number of military, political and scientific men, the botanists, John Bartram and sons, Humphrey Marshall, Manasseh Cutler and Dr. James Greenway of Virginia, who made valuable collections.

Thomas Walter, the author of the Flora Caroliniana, published in London, 1787, was born, 1740, in Hampshire in England. He made his collections on a small area of scarcely more than twenty-five square miles on the Santee river, in South Carolina, but though he declares his collection very incomplete, it contains over a thousand species. He is a most modest man and not an over hasty species-maker. Whenever he is in doubt about a species he does not name it, but calls it "anonymous," for only few, he says in the preface, are allowed to name, and so he concedes to those who are the leaders in science, the right to name those plants now first described. To his name on the title he appends "agricola" (farmer), a learned farmer indeed, as the

whole book is written in Latin.   In front of the book is a copper-
plate (representing *Magnolia frazeri* Walt.) with the inscription :
" To Thomas Walter, Esq., this plate of the new auriculated
Magnolia is presented as a testimony of gratitude and esteem by
his much obliged humble servant, John Fraser."

This man, John Fraser, was a Scotch botanist who collected
from 1780 to 1784 in New Foundland, and from 1785 to 1796 in
the United States.   After a voyage in Russia he came again to
America with his eldest son, John, in 1799.   He visited the Alleghanies, where before, in 1789, he had traveled with Michaux,
and on the summit of the Great Roan it was that he discovered
the beautiful *Rhododendron catawbiense,* now cultivated in many
varieties.   After a visit on the island of Cuba, where he met
Humboldt and Bonpland, father and son returned to England
in 1802.   Once more, 1807, both came to North America.   The
elder Fraser died, 1811, in Glasgow, but his son returned to
America, where he continued his excursions up to 1817.

Palisot de Beauvais (1755–1820) came from St. Domingo to
Philadelphia in 1791.   He extended his excursions to the Cherokee country in the Southern Alleghanies.   Of his collections, a
large part was lost by shipwreck.

L. A. G. Bosc (1759–1828), a friend of Michaux, came from
France to Charleston, in 1798, made some excursions in the
vicinity of that city, traveled northward to Wilmington, N. C.,
and westward to the line of Tennessee.   With a collection of
1600 species, chiefly grasses and cryptogamous plants, he returned
in 1800, to France.   Parts of his collection are found in the herbaria of Ventenat, Martius, Moretti and De Candolle.

Another Frenchman ought to be mentioned here not as a
botanist, but because the matter he treated of is in near relation
to botany.   The Count Volney came to North America as an
exile in 1795, and lived there till 1798.   He had traveled previously in the Orient and had written the famous work, " The
Ruins."   In America he studied the soil and the climate.   His
" Tableau du climat et du sol des Etats Unis d'Amerique," was
published in Paris, 1822.   Though we cannot agree with all he
said, particularly not with the statement that the ancient and
annual fires of the Indians *caused* the prairies, the book contains
much interesting matter.   In the second chapter he describes, in
general, the appearance of the country, particularly the extensive

woodlands, which he brings into three categories, as the southern, the middle and the northern, each characterized by its peculiar trees. Very interesting to Western men is his article on the colony at Vincennes, on the Wabash, and the early French life in Illinois.

The most important collections of this period were made by André Michaux, born in France, 1746. Before Michaux came to this country, he had traveled in Persia, 1782 to 1785. Then, in September, 1785, he embarked for New York, where he arrived in November of the same year. He established two gardens, one in New Jersey, the other near Charleston, S. C., for he was sent by the French government to collect living plants, to be transported to France. His excursions extended from Canada to Florida, and, in the west, to the Mississippi; farther than any collector before him had traveled. From Charleston he started for his first tour to the southern Alleghanies, in April, 1787, and returned the 1st of July; went to Philadelphia and New York, and returned to Charleston in August. Then, in February, 1788, he embarked for St. Augustine, Florida; returned to Charleston, and started again for the Alleghanies. During the following winter he was on the Bahama islands, and brought back to Charleston eight hundred and sixty young trees and shrubs. Then he made several excursions to the Alleghanies of North Carolina, through the valley of Virginia to Maryland and Pennsylvania. From New York he returned to South Carolina, via Baltimore, Richmond and Wilmington, and went again to the mountains. He returned to his nursery with twenty-five hundred young trees, besides many shrubs and other plants. In March, 1792, he sold his nursery near Charleston, and went to Philadelphia, collected in New Jersey and around New York; traveled via Albany and the Champlain lake to Montreal and Quebec, and came back from there to Philadelphia in December.

In July, 1793, he undertook his great journey to the far west; he crossed the Alleghanies of Pennsylvania, descended the Ohio to Louisville; crossed Kentucky and Virginia, back to Philadelphia. In 1794 he visited again the Southern States; in May, 1795, he was in East Tennessee, crossed the Cumberland mountains, arrived, in July, at Louisville, traveled the Wabash up to Vincennes, crossed Illinois, descended the Mississippi in a little boat to the mouth of the Ohio, followed the Cumberland river up to Clarksville, and arrived, via Louisville and North Carolina, at Charleston in August, 1796.

His travels in the Alleghanies are more particularly spoken of by Prof. Asa Gray, in his "Notes of a botanical excursion to the mountains of North Carolina." He had already sent to France more than sixty thousand living woody plants, and forty boxes with seeds, when he returned in 1796. Unfortunately, he suffered ship-wreck off the coast of Holland, but he, and the collections which he brought, were saved, though the latter were damaged. He arrived at Paris in December, 1796, and published, in 1801, his great work on the American oaks, with excellent engravings. He then prepared the material for his Flora Boreali-Americana, but did not live to see it published.

Though he had desired to return to America, he accepted a proposition of Captain Baudin, to take part in an expedition to New Holland, and embarked on the 18th of October, 1801. Arrived in Isle de France, he left the expedition for Madagascar, where a malignant fever caused his death, on the 13th of November, 1802.

Louis Claude Richard arranged the material of his Flora, and, in 1803, it was published by François André Michaux, the son. In this work are described 596 genera (555 vascular and 41 cellular) and 1740 species (1641 vascular and 99 cellular). Though many changes and reductions have been made in the course of time, 17 of the genera, proposed by him as new, are valid yet, and about 350 species.

Considering the vast area he traveled over, often without company, the poor facilities for traveling at that time, the troubles he had to undergo in transporting so many living trees and shrubs, the dangers he had to fear, risking his scalp at every step in the Western wilderness, we must admire that indefatigable traveler. His name stands as a prominent landmark at the dividing line of two periods, from which the labors of working botanists in this country took a new departure.

1800—1840.—Till then, at the close of the eighteenth century, as we have seen, most of the work was done by foreigners, partly engaged by European institutions or by private men. The few Americans did the work at their own expense, for there was no subvention by the Government at that time. Michaux, after his return from Canada, had, in 1792, entered into negotiation with the Government about an expedition across the continent, but without any result. Indeed, the Union, with her four millions of inhabitants, was not a rich country at that time; she could not support scientific pursuits with the same liberality as she, in our times, does with her forty millions. And yet, ten years afterwards, an expedition was sent out under the command of Meriwether Lewis and Daniel Clark, the first that ever crossed the continent.

Before this expedition, only two botanists had visited the Pacific coast, Haenke and Menzies.

Thaddaeus Haenke (1761-1817), botanist with the Spanish expedition under Malaspina, collected about nine thousand species, of which a small part were from Northern California, in 1789. The herbarium is in the National Museum of Bohemia. Haenke did not return to Europe, but settled at Cochabamba, in Bolivia, where he died, 1817. Presl published the description of a part of his collections, under the title: Reliquiæ Haenkeanæ; 2 vols, with 72 plates. Here, by-the-by, may be corrected a geographical error, such as may often be found in learned writings. In Kunth's Enumeratio, vol III, page 361, under Juncus falcatus, we read: a Haenke lectus prope Monte Real (Canada). But Haenke was never in Canada, and it is meant Monterey, in California.

Archibald Menzies (1754-1842), a Scotchman, surgeon in the British marine, came first, 1786, to the North-west coast, and made there some collections. Afterwards, he took part in the famous expedition under the command of Vancouver (1791-95), and visited, several times, San Francisco, Monterey and Nutka Sound, on Vancouver Island. His collections are incorporated in the herbarium of the Edinburg Botanical Society, and, partly, in Hooker's herbarium.

This was all that was known of the botany of the northern Pacific coast, when the first American exploring expedition started from St. Louis, on the 14th of May, 1804. The party ascended the Missouri in keel-boats, cordeled by hand; wintered at Fort Mandan; crossed, in the next season, the Rocky mountains (at the Bitter-root mountains), and descended the Lewis fork and the Columbia river. The botanical collection from the Rocky mountains was, unfortunately, lost; only one hundred and fifty species, collected during the rapid return-march, were saved. These plants were described by Frederick Pursh.

The German botanist, Frederick Pursh, came to North America in 1799, a young man then, but not as young as he is made in Pritzel's Thesaurus, which, on account of so many errors in printing, is, in regard to dates, unreliable. According to Pritzel, he was born in 1794! What a young botanist, crossing the Atlantic, five years old! He was born in 1774.

" My first object after my arrival in America," he says in the preface to his Flora, " was to form an acquaintance with all those interested in the study of botany. Among these I had the pleasure to account one of the earliest, and, ever after, the most valuable, the Rev. Dr. Mühlenberg, of Lankaster, Pa., a gentleman whose industry and zeal for the science can only be surpassed by the accuracy and acuteness of his observations."

Heinrich Ludwig Mühlenberg, born 1756, was a Lutheran preacher in Lancaster, Pa., where he died, 1817. He published a catalogue of North American plants, 1813 (second edition 1818), and a " Description of North American Grasses," 1817.

Then Pursh visited Mr. Humphrey Marshall, already mentioned, the younger John Bartram and his brother William Bartram; Mr. John Lyon, who had the management of Mr. William Hamilton's gardens, and whose successor he was from 1802 to 1805, and Dr. Benjamin S. Barton, Professor of Botany in the University of Pennsylvania, who lived from 1766 to 1815, the author of " Collections for an Essay toward a Materia Medica of the United States," 1798 (second edition 1812-1814); of a " Flora Virginica," first part, 1812, but not continued, and of " Geographical view of trees and shrubs," 1809.

In 1805, Pursh set out for the Alleghanies of Virginia and Maryland; in 1806 he went to the Northern States, as far as New Hampshire; in 1807 he took charge of Professor Hosack's botanical garden of New York; in 1810 he visited the West Indian islands, and returning in 1811, landed in Maine, and embarked the same year in New York for England, where he published his Flora Americæ Septentrionalis, London, 1814, in 2 vols., with 24 engravings. This Flora contains about 740 genera and nearly 3000 species.

It will be easily understood that Pursh's Flora, which was published eleven years after Michaux's, must be richer in genera and species, when we consider that Michaux described only such species as he collected himself, and that Pursh received contributions from many parties. Except the small collection of Lewis and Clark, he used the herbaria of J. Lyon and Barton in Philadelphia, of Hosack in New York, of Le Conte in Georgia, of Peck in Massachusetts, and a number of species received from Alois Enslen, an Austrian gardener, who made large collections in the Southern States and Western territories, which are now in the Imperial Museum of Natural History in Vienna.

Then, in England, Pursh examined the herbaria of Clayton, Pallas, Plukenet, Catesby and Walter. In Bank's herbarium he found a number of the plants collected by Archibald Menzies on the North-west coast.

Pursh returned to America with the intention to explore Canada, where he died in 1820.

C. C. Robin,[1] a Frenchman, traveled, 1802-1806, in Louisiana and West Florida, which at that time included the southern parts of the States of Mississippi and Alabama. The incidents of his voyages he published in 1807, and in an appendix he described the plants, found on his tour, very vaguely, as he indeed was not a botanist. It is not known that he ever brought to France the specimens of those plants. From this written material was fabricated, by Rafinesque, a fancy work called Florula Ludoviciana, published in New York, 1817.

Constantin Samuel Rafinesque-Schmaltz is his full name. He was a Sicilian, and came to America in 1802, where he remained three years, and then again in 1815, and never returned, for he died in 1840, in this country. A. Gray published, in the American Journal of Science and Arts, a paper on his numerous botanical writings. Gray calls him an eccentric but certainly gifted man. It is true, some of his observations are really good, some of his genera and species are acknowledged now and will be in the future, but the greater part are trash; most of his numerous species can never be found, for they have no real existence in nature. He was a polygrapher—he wrote on everything; even poetry, the worst of all, he committed. At last he made a perfect fool of himself; he had such a mania for classification and registration, that he once proposed—twelve new species of thunder and lightning! His travels extended, in 1802-4, over the States of New Jersey, Pennsylvania, Maryland, Delaware and Virginia; in

[1] Not the godfather of the genus Robinia. That was Jean Robin, who lived from 1550 to 1629, in Paris.

1815 and 1816, mostly in New York, New Jersey and Pennsylvania; 1818 to the West (Ohio, Indiana, Kentucky and Illinois). To enumerate all his publications would be a waste of time and paper.

François André Michaux (1770-1855) had already traveled with his father. In 1801 he started again for America, to explore the Western States. In June, 1802, he crossed the Alleghanies of Pennsylvania, on foot, descended the Ohio in a boat from Wheeling to Limestone; crossed Kentucky in a south-westerly direction, and Tennessee as far as Nashville, and returned via Knoxville in East Tennessee, to Charleston, S. C., where he arrived on the 18th of October, 1802.

After his return to France in 1803, he published his " Voyage a l'Ouest des monts Alleghanys," Paris, 1804. The book contains many valuable observations on vegetation, wild as well as cultivated. 1805, he published a work on the naturalization of North American forest-trees, and 1810-1813, his great work, " Histoire des arbres forestières de l'Amerique septentrionale." There is an English translation, published in Philadelphia, 1859; The North American Sylva, three volumes, with 145 plates, uniform with Thomas Nuttall's work with the same title, published 1842-1854, which contains, in three volumes with 121 plates, those trees which are not described in Michaux's Sylva, mostly trees from the Rocky mountains, California and Florida, not known before.

Thomas Nuttall, a native of Yorkshire in England, and a printer by trade, came to America about the year 1808. He was, like Michaux, an indefatigable traveler. In company with John Bradbury, who had already explored the vicinity of St. Louis during the year 1810, he traveled, 1811, the Missouri upward to Fort Mandan; 1816, he was in the Alleghanies, in Kentucky and Ohio. On the 2d of October, 1818, he started from Philadelphia for Pittsburg, descended the Ohio to its mouth, then the Mississippi to the Arkansas river; this river upward to the Fort Smith; from there in a south-westerly direction to the Red river. After his return to Fort Smith, he followed the Arkansas farther up to the mouth of Verdegris river, and Grand river, and northward to the Osage saltworks. This latter excursion was full of hardships, disease, Indian pillaging and peril of life. Returning, he descended the Mississippi to New Orleans, where he arrived on the 18th of February, 1820.

From the manner of writing, we may often perceive the character of a man. Whoever may read his " Journal of travels into the Arkansas territory," published in Philadelphia, 1821, will be delighted at the plain, unpretending style, the " unvarnished tale," as he expresses himself in the preface, and will divine in Thomas Nuttall an amiable man.

In the years 1834 and 1835, Nuttall crossed the Rocky mountains to the Pacific coast, explored Oregon and California, made an excursion to the Sandwich islands, and returned around Cape Horn to the Atlantic coast. Besides the above-mentioned books, he published his "Genera of North American plants," in two volumes, Philadelphia, 1818 ; an "Introduction to systematic and physiological botany," Cambridge, 1827, and numerous descriptions of new plants, mostly in the Proceedings of the Academy of Natural Science, in Philadelphia. He died at the ripe age of seventy-three, on the 10th of September, 1859, in Lancashire in England.

Nuttall and Bradbury are mentioned by W. Irving in his Astoria, in which he describes the voyages of the parties sent out to Oregon by Mr. Astor. As both gentlemen left the expedition on the upper Missouri, these voyages had no further relation to botany.

Several other foreign botanists collected at that time in North America. Alire Raffenau Delile, professor of botany at the University of Montpellier, in France, after his return from the French Scientific Expedition in Egypt, a prominent member of which he was, came over to America and collected during three years, in the vicinity of Wilmington, N. C.

José Francisco Correa de Serra, secretary of the Royal Academy of Lisbon, came in the year 1813 to New York and Philadelphia, from where he made several excursions.

From 1817 to 1823 Mr. Milbert collected for the Museum of Natural History at Paris. He lived in New York and extended his excursions to the Ohio, Mississippi, Lake Superior and Canada.

Active American botanists of that time were Amos Eaton, professor in Albany, N. Y. He lived from 1776 to 1842, and published the first edition of his Manual of Botany, 1817, and of eight editions the last in 1841.

James Bigelow was professor of botany in Boston. His first edition of Florula Bostoniensis appeared in 1814, the third in 1840, and his American Medical Botany, 1817–1821, in three volumes with sixty colored plates.

William Baldwin, born in Pennsylvania 1779, collected in 1811, in Delaware, then in Georgia and Florida, and went, 1817, to Buenos Ayres, in South America. He died on the 31st of August, 1819, as a member of Major Long's first expedition, in Missouri. Darlington published, 1843, Reliquiæ Baldwinianæ. This expedition, by order of the Government, under the command of Major Long, started from Pittsburg in April, 1819, and proceeded the same year up the Missouri to Council Bluffs, where they wintered. Dr. Baldwin, the botanist of the expedition, sick already when the party set out from Pittsburg, died in Franklin, Mo., and Edwin James took his place, who compiled the account of the expedition in two volumes, 1823; the same year it was published in London in three volumes. The party started again on the 6th of June, 1820, from Council Bluffs, moved up the Platte river and examined the mountains from the South fork of the Platte to the Arkansas. Dr. James ascended the grand peak described by Major Pike in an account of his expedition in the years 1805–1807, which furnished no botanical matter. This peak is, in the narrative, called James' Peak; Fremont afterwards changed it to Pike's Peak, although Pike had only seen it and James was the first that ascended it. In returning, one part of the command followed the Arkansas river, the other the Canadian river. The catalogue of the collected plants, 500 to 600 species, was published by James, in 1825, in the Transactions of the American Philosophical Society, II, 172–190. James died 1861, near Burlington, in Iowa; he was born in Vermont, 1797.

In the period from 1820 to 1830 several Floræ of more or less limited parts of the United States were published. The best known are: Botany of South Carolina and Georgia, 1821-1824, in two volumes, with twelve plates, by Stephen Elliot, professor in Charleston, where he died 1830; the Flora of the Northern and Middle States, by John Torrey, 1824, and the Flora Cestrica (of Chester county, Pa.), 1826, by William Darlington, who lived from 1782 to 1863, and published the third edition of his Flora in 1853.

Lewis David v. Schweinitz, born in 1780 at Bethlehem, Pa., where he lived to 1834, published, in 1821, "Specimen Floræ

Americæ Septentrionalis Cryptogamicæ," containing the liverworts, and 1825, a Monograph of the genus Carex. He collected the fungi of Carolina, a catalogue of which, containing
1373 species, was published by Schwægrichen, the well known
cryptogamist, in Leipzig, 1822.

The catalogue of plants collected in the North-western territory during Major Long's second expedition, is written by
Schweinitz. This expedition was described by Wm. H. Keating,
the geologist of the party, and published in London in 1825, in
two volumes, entitled, " Narrative of an Expedition to the Source
of St. Peter's river, Lake Winnepeek, Lake of the Woods, etc.,
performed in the year 1823." The party left Philadelphia on the
30th of April, and the route of the expedition was the following:
Wheeling, Fort Wayne, Chicago, Fort Crawford, Fort St. An-
thony, up the St. Peter's river to its source, down the Red
river to Lake Winnipeg, Rainy lake, Fort Williams on Lake
Superior and return on the lakes. Edwin James was appointed
botanist, but he missed the place of meeting, so Mr. Thomas Say,
the zoölogist of the expedition, undertook to collect the plants.
As several boxes containing collections, and dispatched during
the expedition, were lost, the botanical collection was very poor,
only 130 species. As poor as the collection was, the description
of the species called new by Schweinitz, are mostly riddles not
yet solved. So the expedition, otherwise interesting, was unim-
portant as to botany.

Dr. Douglass Houghton, who met, in 1845, a sad end by
drowning in Lake Superior, was a member of Schoolcraft's Expedition to the sources of the Mississippi river in 1832, the narrative of which was published in 1855. He collected about 250
species of plants, only a few of which were new.

Up to this time a number of botanists were at work on the
Pacific coast.

Adelbert Chamisso de Boncourt (1781–1838) and Frederik
Eschscholtz (1793–1831) were commissioned by the Russian
government, the one as naturalist, the other as physician, to
the Russian Exploring Expedition in the Pacific and Behring
straits, under the command of Kotzebue, 1815–1818, and explored
Alaska and a part of the coast of California. The collections are
partly in the royal herbarium at Berlin, partly in St. Petersburg.

The plants collected by Lay and Collie, of the expedition of

Capt. Beechey to the Pacific, 1825–1828, were described by Sir Wm. Jackson Hooker and G. A. Walker-Arnott, and published in London, 1841, a quarto volume with ninety-four plates. A part of these plants were collected in California.

Carl Heinrich Mertens, born in Bremen, 1796, took part in the Russian expedition under the command of Capt. Lütke, 1826–1829. Amongst his collections was a number of plants from the Island of Sitka, which, as Mertens not long after his return died in St. Petersburg, were described by Bongard in the memoirs of the Academy of Science of St. Petersburg, 1832. His account of the vegetation of Sitka was already published, 1827, in Berlin, by A. Chamisso, with observations of the same.

A member of the same expedition was F. H. v. Kittlitz, who published twenty-four fine landscape views of the Pacific islands and coasts, amongst which are four that give a good idea of the vegetative character of Alaska. Three of them are rather roughly copied in the U. S. Agricultural Report for 1868.

David Douglas, born in Scotland, 1790, traveled for the Horticultural Society of London. He arrived in July, 1823, in New York, made excursions through New York State to Canada, and returned to London 1824. The society was so well pleased with his collection that he was sent the same year to Oregon, where he arrived in February, 1825, on the same ship with Dr. John Scouler, another Scotchman, who was afterwards professor in Dublin, and died, seventy-two years old, in 1871, in Glasgow, his birthplace. Douglas took up his residence in Fort Vancouver, and made from there excursions into the interior of Oregon Territory, and to Northern California, then he crossed the Rocky mountains to the Athabasca river, and to York Factory on the shore of Hudson's bay. In October, 1827, he returned to England. In 1830 he undertook his second voyage to Oregon and Upper California, and in 1833 he crossed the Pacific to the Sandwich islands, where he lost his life in a horrible manner. It was on the 12th of July, 1834, that on an excursion he fell into a deep excavation made for the purpose of capturing wild beasts; a wild ox, plunging soon after him into the same hole, killed him. This was a time of disaster for traveling botanists. A year before Douglas, in February, 1833, Thomas Drummond died in Havana; two months only after Douglas, was Carl Beyrich taken away by the cholera at Fort Gibson, the next year Joseph Frank died in New

Orleans, and in 1837 H. B. Croom lost his life by shipwreck on the coast of North Carolina.

H. B. Croom was born in North Carolina, 1799; his catalogue of plants of Newbern, N. C., was, after his death, published by Torrey.

Joseph Frank came to America, 1835, and collected for a botanical society in Germany (the Unio Itineraria). He traveled in Pennsylvania, Ohio, Missouri and Louisiana.

Carl Beyrich, another German, was sent by the Prussian government. He collected, 1833, in North and South Carolina and Georgia 1300 species in one season. The next year he went with a military expedition (probably that of Col. Dodge) from St Louis to the Indian Territory, to leave it no more.

Thomas Drummond, brother of the well known Australian traveler, James Drummond, took part in Franklin's second expedition as an assistant of Dr. Richardson, in 1825. At Cumberland House he left the party to explore the Rocky mountains of the British Territory. In 1831 he collected in the Alleghanies, and then in the vicinity of St. Louis and New Orleans, where he embarked for Texas. He explored the country around Austin, Brazoria and Galveston, went to Apalachicola, in Florida, and started from there in February, 1833, for Havana, where he died in the month of March.

Already before Drummond, in 1827–1830 Texas was, in its more western parts, explored by Jean Louis Berlandier, from Geneva. He also, though later in 1851, died far from home in Matamoras, on the Rio Grande.

Maximilian, prince of Wied, traveled in the Western Territories in the years 1832–1834, and brought back to Germany a small collection of about two hundred species, which were published by Nees v. Esenbeck, professor of botany in Breslau. There was nothing new except the genus Sarcobatus, proposed by Nees and afterwards described again by Torrey under the name of Fremontia.

Many American botanists were at work in this period, collecting the plants about their homes or exploring the vegetation of larger districts. The most prominent ought to be named here: In Massachusetts, Bigelow, Tuckerman, Oakes, Dewey; in Connecticut, Barratt; in New York, Sartwell, Carey, Beck, Bailey; in Pennsylvania, Pickering, Durand and Darlington; in North

Carolina, M. O. Curtis; in Georgia, Boykin and Le Conte; in
Florida, Chapman, Leavenworth and Blodgett; in Alabama,
Gates; in Louisiana, Hale, Carpenter and Riddell, who, in 1835,
published a Flora of the Western States, and afterward a Flora
Ludoviciana.   Other catalogues of local floras were compiled of
the plants in the vicinity of Charleston, S. C., by Bachman, 1834;
of the plants of Columbia, S. C., by Gibbes, 1835; of the plants
near Baltimore, by Aikin, 1836.   Dr. Pitcher collected in Arkansas, afterwards in Michigan; in Kentucky Dr. Peter and Prof.
Short; in Illinois the same and Buckley; in Ohio, Lea and Sullivant; in Michigan, Wright; in Wisconsin, Lapham; in Missouri,
Engelmann.

In the botany of the Californian survey, the first of which
volume is now published, we find often the quotation of Plantæ
Hartwegianæ.

Theodor Hartweg, a German gardener, was sent by the English
horticultural society to Mexico and California, where he collected
during the years 1838-1839.   Many of the plants described and
published by G. Bentham, under the above-mentioned title, in
1839-1848, occur in what was formerly the northern Mexican
countries.

J. N. Nicollet, a Frenchman, employed under the Bureau of Topographical Engineers (since 1838), explored the basin of the
upper Mississippi during the years 1836-1840.   To his party was
attached the German botanist, Carl Geyer, the large collections of
whom were sent to Drs. Torrey and Gray.   These two bright
stars had already risen above the horizon of the botanical firmament, and commenced, at the close of the fourth decennium, their
great work, the Flora of North America, opening a new epoch in
the history of American botany.

The interest in the science of botany was now wide awake
amongst the American public, and the Government bore its rich
share of it, spending large sums for scientific purposes, by attach-
ing scientific men to the nearly unbroken series of expeditions
and surveys which were now undertaken.

HISTORICAL SKETCH
OF THE SCIENCE OF BOTANY
IN NORTH AMERICA FROM 1840 TO 1858

Frederick Brendel

HISTORICAL SKETCH OF THE SCIENCE OF BOTANY IN NORTH AMERICA FROM 1840 TO 1858.

BY FREDERICK BRENDEL.

[*Continued from p. 771, Vol. XIII, American Naturalist.*]

AT the time when Torrey and Gray commenced their first work on the Flora of our continent north of Mexico, Sir William Jackson Hooker, the celebrated English botanist, had finished his great work on the Flora of British America, two volumes, in quarto, with 238 plates, London, 1833-1840. But, before we proceed farther, we have to review the early history of botany in the most northern and Arctic regions of North America.

Hans Egede, a Danish missionary, was, from 1721 to 1736, in Greenland. After his return to Denmark, he published, in 1741 a description of that country. He describes, vaguely, some plants, with some drawings on one plate, but it is rather difficult to make out what the drawings mean. Afterwards his son, P. Egede, made some botanical collections, which, as well as those of Giesecke, who published a Flora Grœnlandica, 1816, in Brewster's Edinburg Encyclopædia, and those of Wormskiold, are preserved in the herbaria of Hornemann and M. Vahl.

Some Greenland plants were described, 1770, by Rottboell, professor in Copenhagen, and, in the same year a history of Greenland was published by the missionary Cranz; the plants in it were described by Schreber.

The largest collections were made in this century by Jens Vahl, the librarian of the botanical garden in Copenhagen, who traveled nine years in Greenland, and probably there will not many new discoveries be made. J. Lange's catalogue of Greenland plants (in Rink's work on Greenland, 1857), contains 320 species in 52 orders.

A list of plants collected on the coast of Baffin's bay was published by Robert Brown, in 1819, and by the same a "Chloris Melvilleana," 1823, containing 131 species, of which 80 are phenogamous, collected at different times by Sabine, Edwards, Ross, Parry, Fisher and Beverley.

Scoresby's collection in East Greenland, was described by Hooker in 1823, and that of Sabine in 1824.

John Richardson, born in Scotland, 1787 (died 1865), was the naturalist of the expedition from the shores of Hudson's bay to the Polar sea, 1823, under the command of Franklin. This expedition started from York Factory, on Hudson's bay, and proceeded via Cumberland House, Carlton House, Fort Chippeway, on the Athabaska lake, Fort Providence, on the Slave lake, and Fort Enterprise, 65° N. latitude, to the Coppermine river, then along the coast eastward to Cape Turnagain, the Hood river up to Fort Enterprise, to Norway House, on the Winnipeg lake, and back to York Factory. The collection of plants contained 700 species, and was published by Richardson in the botanical appendix to Franklin's Narrative, printed 1823.

The narrative of the discoveries on the north coast of America, by Simpson and Dease, in 1837, published in 1843, contains a catalogue of plants examined by Sir William Hooker, but nothing new; all the species were already collected by Richardson.

Berthold Seemann (born in Hanover, 1825), the naturalist on board of H. M. S. *Herald*, under the command of Captain H. Kellet, during the years 1845-1851, described, in a letter addressed to Sir William Hooker (in Journal of Botany), the arctic Flora of Kotzebue Sound, and published, 1852-1857, the botany of the expedition, the first part of which contains the Flora of Western Esquimaux land.

Bachelot de la Pyiaie, a French botanist, explored, in 1819 and

1820, Newfoundland and the little islands of Miquelon and St Pierre. He published, 1829, Flore de l'isle de Terre Neuve, which was not finished, and contains only a description of cryptogamous plants.

A Flora of Labrador was compiled by E. Meyer, professor of botany in Kœnigsberg, in 1830, from a small collection by the missionary Herzberg, and a number of species made known by Schrank, professor of botany in Munich. These plants were collected by a Danish missionary, Kohlmeister, probably the same that Pursh calls Colmaster in his Flora, and the plants of which he found in the herbaria of Dickson and Banks. The number of all the species of Labrador known at that time, was 198, of which 30 are cryptogamous.

The north-west coast was visited, 1838, by the expedition of H. M. S. *Sulphur*, under the command of Sir Edward Belcher. This expedition explored the Pacific during the years 1836-1842. The botanist was Mr. Barclay, in the service of the Kew garden, assisted by the surgeon Hinds and Dr. Sinclair. The parts visited were Prince William's sound, Port Mulgrave, both under 60° N. L., Sitka, Nutka sound, San Francisco, Sacramento river and Monterey in California. The botanical collections were described by George Bentham, in "Botany of the voyage of H. M. S. *Sulphur*," 1844, with 60 plates.

The U. S. Naval exploring expedition, under the command of Charles Wilkes, which crossed the Pacific during the years 1838 to 1842, in every direction, arrived, 1841, in Oregon. Charles Pickering was collector on this expedition. The Columbia river up to Walla Walla, and the Willamette valley were examined: afterwards the Sacramento river down to San Francisco. In Oregon were collected 1218 species, and 519 in Northern California; the whole collection of this expedition amounting to 9600 species, were examined. The phanerogamous plants were described by Dr. Torrey; the ferns of the expedition, by Dr. Brackenridge; the mosses by Mr. Sullivant; and the lower cryptogams by other botanists.

N. J. Andersson, a Swedish botanist, naturalist of the voyage around the world of the Swedish frigate *Eugenie*, collected in 1852, in California; he took particular notice of the willows, and in 1858, he published in Proceedings of American Academy of Arts and Sciences, a "Synopsis of North American Willows," of which he enumerates fifty-nine species, a number of which he degraded to the rank of sub-species in his Monographia Salicum, 1863. He is the author of the genus Salix in Decandolle's Prodromus.

Besides the Rocky mountains and California, another large field opened for exact exploration. The Mexican war and the acquisition of new territories caused a long series of expeditions to California and those tracts of land which form with West Texas the North Mexican botanical province.

In June, 1842, Lieut. Fremont set out from the mouth of the Kansas river, followed that river about one hundred miles, passed over to the Platte river, traveled up the river to the junction of the north and south fork, where the party divided, one part following the north fork to Fort Laramie, the other proceeding along the south fork to Fort St. Vrain, and from there to Fort Laramie. Then the expedition followed the north fork and the Sweetwater river up to South Pass, and the Wind River mountains, the highest peak of which, afterwards called Fremont's Peak, he ascended. Returning, the Platte river was followed to its mouth. The collection of plants, consisting of 352 species, contained fifteen new ones, described by Torrey.

The collections of Fremont's second expedition, during the years 1843 and 1844, which extended to Oregon and California, were greatly damaged, so that in many instances it was extremely difficult to determine the plants. Torrey furnished the description of a few new genera and species, which, with four plates, was published in App. C. to Fremont's Report. One of these new genera he named Fremontia, but this name was afterwards withdrawn, as Nees had already described the plant under the name Sarcobatus, and Fremont's name was transferred to another new Californian genus of the order of Sterculiaceæ.

Two other expeditions were undertaken by Fremont in 1845-1846 and 1848, extending to California. Large collections were made again, but the greater part of them were destroyed by the same mishaps. Some of the new genera that were saved for examination were described and published in 1850, by Professor Torrey in the Smithsonion Contributions, as " Plantæ Fremontianæ," with ten beautiful plates.

Emory's military expedition traversed in June and July, 1846, the plains from Fort Leavenworth to the bend of the Arkansas, followed this river to the Pawnee fork, crossed the Raton moun-

tains (7000 feet) and the ridge between the Canadian river and
Rio Grande to Santa Fé; then again the dividing ridge (6000
feet) between the Rio Grande and Gila, followed the latter to the
Colorado of the West, and arrived at St. Diego. The botanical
collections (about 200 species) were examined by Torrey, the
Cactaceæ by Dr. Engelmann, and published in Appendix 2 of
Emory's Report. A small number of plants was collected by
Lieut. Abert, amongst which was nothing new.

Dr. A. Wislizenus, born in Germany, 1810, left St. Louis in the
spring of 1846, with the intention of traveling in North Mexico
and Upper California. He undertook the journey at his own expense, and war was not yet declared, when he arrived at Chihuahua; but there he was arrested as a spy, and transported to Cosi-
huirachi, at which place he was left in a " passive " condition; that
means as to his free will to leave; for, as a collector, he was very
active on this rich field, where he collected so many species not
found before. Six months afterwards, Colonel Doniphan's troops
occupied that part of the country, and Wislizenus accepted a
situation in the medical department of the American army, and,
instead of going westward as he first intended, he followed the
army to Monterey, and returned via Matamoras to the States.
He collected a large number of plants. In an Appendix to the
" Memoir of a tour to North Mexico in 1846 and 1847, by A. Wislizenus, M. D., printed for the use of the Senate of U. S.," the
botany of the explored country is described by Dr. Engelmann.
Amongst the new species were over thirty new species of Cactus.

West Texas was extensively explored since 1835, when Ferdi-
nand Lindheimer (born in Germany, 1802), settled at New Braun-
fels, where he lives yet. His large collections were named and
described by Gray and Engelmann, in Boston Journal of Natural
History, as Plantæ Lindheimerianæ, part I in 1845, part II in
1847. Many of these plants were shortly afterwards described
by Scheele, in Linnæa, from a collection brought to Germany by
the geologist F. Roemer, who studied the geology of Texas in
1846-1847, and received many specimens from Lindheimer.

Completed and extended to the whole area of the Rio Grande,
were these explorations by Gregg and Wright.

Josiah Gregg, the author of the " Commerce of the Prairies,
1844," made large botanical collections, but died soon (1849) in
California.

Charles Wright spent several years in Texas, the botany of which country he studied. Then, in 1849, he went westward to El Paso, in New Mexico. His rich collections of plants were placed in the hands of Prof. Gray, who described and published " Plantæ Wrightianæ " in the third volume of Smithsonian Contributions, with ten plates. In 1851–1852, he was again in New Mexico, the collection of which tour furnished the material to the second part of Plantæ Wrightianæ, with four plates, in the fifth volume of the Smithsonian Contributions.

Another well-known botanist explored New Mexico at the same time. August Fendler came, about the year 1836, from Germany to North America. In 1846 he left Fort Leavenworth with a military train, followed the Arkansas river up to Fort Bent, crossed the mountains to Santa Fé, where he made his principal collections from April to August, 1847. An account of his collection Prof. Gray published in the Memoirs of the American Academy, Vol. IV, under the title, Plantæ Fendlerianæ. Fendler resided a long time near St. Louis; went afterwards to Venezuela, and is now collecting on the Island of Trinidad.

Dr. Woodhouse was a member of the expedition down the Zuñi and Colorado rivers under the command of Capt. Sitgreaves, in 1850. His collections, placed for examination in the hands of Prof. Torrey, consist of three portions: the first, collected between Neosho and Arkansas rivers, and on the north fork of the Canadian, and the second, from Texas, contain nothing new. The plants of the third portion were collected in Arizona and California. The catalogue of the latter (about 180 species) is published with twenty-one plates, in 1853, with Sitgreaves' Report. There is described a new genus of the order of Amarantaceæ and several new species.

In the year 1852 an expedition under the command of Capt. Marcy explored the Red river to its sources. The botanical collection of 200 species, made by Dr. Shumard, was examined by Prof. Torrey. and published in App. G of Marcy's Report with twenty plates.

The botany of these formerly Mexican provinces was nearly unknown before Berlandier; but by the collections of the above-named botanists much light was thrown upon it; their work was further advanced by the Pacific railroad explorations and the Mexican boundary survey, and will be completed by the surveys of the

Territories in progress yet. Here may be mentioned, though not in the compass of this historical sketch, the surveys of Hayden, Powell, Wheeler, and King, who surveyed the State of Nevada, and whose report contains, in the fifth volume, the botany of Nevada and Utah, by Sereno Watson.

Several expeditions were ordered by the Government, in 1853, to cross the Rocky mountains to the Pacific, along certain parallels, to explore the most practical route for a railroad to the Pacific coast. The parties consisted of a military command and a number of technic and scientific assistants. The reports of these explorers were afterwards published by the Government in thirteen quarto volumes, extensively illustrated and full of the most important scientific matter. The botanical collectors were the following: Dr. Suckley, naturalist to the party of Gov. Stevens, who proceeded between the parallels 47 and 49 to Oregon. The botanical report of this route, with six plates, we find in the last volume; it contains the collections of Suckley on the plains, 323 species, examined by Prof. Gray (one genus and three species were new), and a catalogue of plants from Washington territory collected by Dr. Cooper. Those east of the Cascade range, 75 species, of which two were new, were examined by Prof. Gray, those of the west-side, 386 species, of which one was new, collected by Suckley and Gibbs, were examined by Torrey and Gray. A general report on the botany of the route is written by Dr. Cooper.

In Vol. II, Torrey and Gray reported on the collection of plants made by F. Creutzfeldt, a German gardener from St. Louis, who was engaged as botanist under the command of Capt. Gunnison, and was killed with the same, by the Indians, near Sevier lake, Utah. He collected 124 species, with two new ones; the report is illustrated by three plates. After the murder of Gunnison, the party reached, under Lieut. Beckwith, the Great Salt lake, where the winter was passed. J. A. Snyder, the topographer of the party, took charge of the botanical collections made along the 41st parallel, from the Great Salt lake to the Sacramento river. The plants, 59 species, were published by Torrey and Gray. There were seven new species, illustrated on seven plates.

The richest collection was that of Dr. J. M. Bigelow, under Capt. Whipple, along the 35th parallel; it is published in Vol. IV, and contains 1109 species of vascular plants, amongst which nine genera and seventy-two species were new, illustrated by twenty-five plates. In a separate report Dr. Engelmann described fifty Cactaceæ, of which eighteen were new and illustrated by twenty-four plates, and Sullivant seventy-two mosses, with twelve new ones and ten plates. Very instructive is the general description of the soil, of the productions along the route, and the forest trees by Bigelow.

The route near the 32d parallel, from El Paso to Preston on the Red river, was explored by Capt. Pope. Dr. Diffendorfer made the botanical collections, which contained 268 species, of which three genera and thirteen species were new. The catalogue is published by Torrey and Gray in Vol. II of the Reports, with ten plates.

Dr. A. L. Heermann was the naturalist under the command of Lieut. Williamson, who explored the passes in the Sierra Nevada and the coast range. The catalogue of eighty-eight species, amongst which were fourteen new ones, with eighteen plates, described by Durand and Hilgard, we find in Vol. v, with a sepa-.rate collection of the geologist, W. P. Blake, containing eighty-seven species, with six new ones and ten plates, described by Dr. Torrey.

Vol. VI contains the interesting botanical report of Dr. J. S. Newberry, geologist under the command of Lieut. Williamson when he explored the country between the Columbia river and Sacramento river. Besides an article on geographical botany, he described the forest trees with ten plates, and added a catalogue of 531 vascular plants (two genera and eight species were new) with six plates, twenty mosses and ten lichens. He was afterwards a member of the expedition on the Colorado of the West, under the command of Lieut. Ives, in 1857–1858. The report was published in 1861, and part 4 contains the catalogue of plants examined by Gray, Torrey, Engelmann and Thurber, 400 species with ten new ones.

Dr. Thomas Antisell collected, under the command of Lieut. J. G. Parke, between the Rio Grande and Southern California, 281 species (one genus and three species were new), which are published with eight plates in Vol. VII of the reports.

The Mexican boundary survey began in 1849, and continued with an interruption, and after a reorganization under Major Emory in 1853, till 1856. The report of Emory was published,

1858, in two large volumes; the first half of the second volume contains the botany. An introductory chapter on geographical distribution and botanical features of the country was written by Dr. C. C. Parry, the catalogue of plants, with descriptions of twelve new genera and 195 new species, with illustrations on sixty-one plates, by Torrey, partly by Gray and Engelmann, who elaborated the Cactaceæ separately and described ninety-two species, of which not less than forty were new, with seventy-five plates. The whole work contains under 2140 species, 235 new ones. The most new species, besides the Cactaceæ, we find amongst the Euphorbiaceæ (36), described by Engelmann, then under the Compositæ (32), and the Scrophulariaceæ (19), both orders described by Prof. Gray. Eight orders comprise half the species of the collection: Compositæ 430, Leguminosæ 212, Euphorbiaceæ 101, Cactaceæ 92, Scrophulariaceæ 71, Cyperaceæ 61, Labiatæ 53 and Cruciferæ 47. The large order of Gramineæ, elaborated by Dr. George Thurber, was unfortunately omitted on account of the already too great size of the volume. Geo. Thurber was one of the botanists of the survey under Bartlett at the same time with Dr. J. M. Bigelow. ·Gray published already, in 1854, in Memoirs of the American Academy of Arts and Sciences, N. S. Vol. v, " Plantæ novæ Thurberianæ," twenty-eight species, of which six belong to six new genera. Charles Wright was attached by Col. Graham to his separate corps of the survey. Under Emory, Dr. C. C. Parry and A. Schott made the botanical collections.

The important result of all these explorations was not only the multitude of new genera and species made known, but the light thrown upon the distribution of North American plants. It was recognized that there is an unmistakable difference between the eastern wooded, the central treeless and the Californian sections of temperate North America, of which the first may be called the sylvan, the second, the campestrian, and the third, the Californian botanical province. The campestrian province reaches from West Texas to Southern California, and far north on both sides of the Rocky mountains; the Sierra Nevada and Cascade range, in Oregon, form the western border, but on the east side there is no sharp line, the prairies stretching into the wooded country. That the flora of East Texas is identical with that of Louisiana and the other Gulf States, Engelmann has sufficiently demonstrated in the Proceedings of the American Association, fifth meeting, 1851.

At the time of Fremont's first expedition, two German botanists directed their lonely ways to the Rocky mountains and to the Territory of Oregon.

Carl Geyer, born 1809, came to America in 1835. As already mentioned he was employed as a collector by Nicollet, afterwards he crossed the Rocky mountains under 40° N. L. to Oregon territory. His rich collections were sent to Sir William Hooker, who examined the plants and described fourteen new species in *Journal of Botany*, 1845 to 1856. Geyer himself furnished interesting remarks on the features of the country. He returned in 1845 to Germany, and died there in 1853.

Lueders, from Hamburg, crossed the Rocky mountains in 1843, and made collections in Oregon Territory. Fremont met him near the Cascades of the Columbia river, where he (Lueders) lost his package by capsizing of his canoe in the rapids, an accident which Fremont memorized by naming a little bay in that locality after his name Lueders' bay, probably a poor reparation for his loss. Nothing was known of him afterward.

Descriptions of plants collected by Dr. Wm. Gambel in the Rocky mountains and California were published by Thomas Nuttall in the Proceedings of the Academy of Natural Sciences of Philadelphia, in 1848. He proposed twelve new genera and 106 new species, but not all of these remained valid.

Captain Stansbury explored, 1849–1850, the valley of the Great Salt lake. His report was published by the Government in 1852. Appendix D contains the botany examined by Torrey, 114 species, of which three were new and some new varieties, with nine plates.

The knowledge of the botany of British America was greatly promoted by Sir John Richardson's Arctic searching expedition, published in two volumes, London, 1851. The object of the expedition was the search for the lost Captain Franklin along the north coast. The voyage was made in boats from Lake Superior via Lake Winnipeg to the Mackenzie river, then from Great Bear lake to the Coppermine river and lasted from May, 1848, to Sept. 1849. In the Appendix (more than half the second volume) we find chapters on the physical geography, climatology and geographical distribution of plants north of the 49th parallel, with most valuable observations. There is a list of trees and shrubs with their northern limits, and a table of the distribution of Carices, which was prepared by Dr. Francis Boott, one of the best authorities and author of the beautiful " Illustrations of the genus Carex," the 4th part of which was after his death published by J. D. Hooker. Boott was born in Boston, 1792, and died in London, 1863.

In the summer of 1848, Prof. Agassiz made a scientific excursion to the Lake Superior with a number of students. He published, 1850, a volume on the physical character, vegetation and animals. Two chapters treat of the botany on the shore of Lake Superior compared with that of the Jura and the Alps. The accounts of such excursions are highly interesting, when related by competent botanists, *e. g.*, that published by Prof. Gray in 1841, in *Silliman's Journal*, " Notes of a Botanical Excursion to the Mountains of North Carolina."

The most prominent American botanists of our times are Torrey, Gray and Engelmann.

John Torrey was born in New York, 1798, and died on the 10th of March, 1873. Author of many botanical writings, he published, as early as 1819, a " Catalogue of plants growing spontaneously within thirty miles of the city of New York ;" in 1824 a " Flora of the northern and middle sections of U. S.," of which only Vol. 1 was printed, containing Classes I-XII of the Linnæan system, which was at that time yet in general use; 1826, a " Compendium of the Flora of the Northern and Middle States ;" 1836, a " Monograph of the North American Cyperaceæ" (in Annals of the Lyceum of New York, Vol. III) ; from 1838 to 1843, with Asa Gray, the first two volumes of the Flora of North America already mentioned. It contains the orders from Ranunculaceæ to Compositæ, and was not continued at that time, but will be finished now, since the large amount of new material brought from the Western explorations is nearly worked up.   In 1843 appeared his " Flora of the State of New York," two large quarto volumes, with 162 tables, forming the second part of the Natural History of New York. In the preface we find a historical sketch of the botanists and their work in the State before that time.   His other writings are already mentioned.

Asa Gray was born on the 18th of November, 1810, in Paris, Oneida county, New York, and is now Professor of Botany of

Harvard University, at Cambridge, Mass. His first work was published 1834-1835, " North American Gramineæ and Cyperaceæ," two volumes, containing each one hundred species, illustrated by dried specimens. It was followed by " Elements of Botany," 1836; " Melanthacearum Americæ septentrionalis revisio," 1837; the " Botanical Textbook," 1842 (third edition, 1850); "Chloris Americana," illustrations of new, rare and otherwise interesting North American plants. Decade I, with ten beautiful plates, 1846 (not continued). The first edition of his well-known " Manual of the Botany of the Northern States," appeared in 1848, and was followed by many editions. " Genera Floræ Americæ boreali-orientalis illustrata, Vol. I and II, with 186 tables," from Ranunculaceæ to Terebinthaceæ, was not continued. Already mentioned are many of his contributions in public documents, Smithsonian publications and scientific periodicals, too numerous to be all named, but all of the greatest value.

George Engelmann, born in 1810, in Frankfurt-on-the-Main, came to America about the year 1834, and has resided since that time in St. Louis, Missouri. Except his writings mentioned above, he has published, in different periodicals, a number of monographs of difficult orders and genera, *e. g.*, Cactaceæ, 1856, Cuscutæ, 1859. His papers on North American Juncus, Quercus, Yucca, some Coniferæ and Gentianæ are later.

Other active botanists of that period are A. Wood, who published a " Classbook of Botany," which is much in use. There is a " Botany of the Northern States," by L. C. Beck, professor in Albany, who lived from 1798 to 1853; a " North American Botany," by Eaton and Wright; an " Introduction to Botany," by Comstock; an "American Flora," by Strong; a " Botany of the Southern States," by Darby.

Local floras and catalogues of plants were compiled, by Dewey: Report on the Herbaceous Plants of Massachusetts, 1840; by Emerson: Report on the Trees and Shrubs of Massachusetts, with seventeen plates, 1846; by Lapham: Plants of Wisconsin, 1849, and a Catalogue of Plants of Illinois, published in the second volume of Trans. of Ill. State Agric. Soc., 1857. Catalogues and notes on the botany of this State were published previously, 1826, by Dr. L. C. Beck, and, 1843, by C. Geyer, with notes by Dr. Engelmann, both in *Silliman's Journal*, then 1845, by Dr. C. W. Short in the *Western Journal of Medicine*. Much has been done for the knowledge of the botany of Illinois by Dr. S. B. Mead, of Augusta, Hancock county, and Dr. G. Vasey, of McHenry county.

Amongst the American botanists, although born in Germany, may be named Dr. Rugel. He came, in 1842, to America, and settled afterwards in East Tennessee, where he lately died. He collected in the South-eastern States, and used to send his collections to Shuttleworth, in Geneva (Switzerland).

There is a number of catalogues which fall partly in the latter time of the second period, that of Bachman, of the plants in the vicinity of Charleston, S. C., 1834; by Gibbes, of the plants of Columbia, S. C., 1835; by Aiken, of the plants near Baltimore, 1836; by Lea, of plants collected in the vicinity of Cincinnati, after his death published by Sullivant. The fungi of the collection were examined by Berkeley.

Ravenel published a paper on the plants of the Santee canal, 1850, and Kirtland one on the climate, flora and fauna of the southern shore of Lake Erie, 1852.

Publications on single orders exist of Jos. Barratt, " Salices Americanæ " and " North American Carices," 1840; of Tuckerman, " North American Lichens," 1848; of Sullivant, " Musci Alleghaniensis," 1846, and " Bryology and Hepatology of North America," 1847; of Bailey, on "North American Algæ." 1848; of Curtis, on " North American Fungi," 1848, both in *Silliman's Journal;* of Dewey, " North American Carices," in *Silliman's Journal;* of Sartwell, " Carices Americæ septentrionalis exsiccatæ," 1848–1850 (158 species); of Alex. Braun (professor in Berlin, who died lately), " Equisetæ and Charæ," in *Silliman's Journal;* of Harvey (professor in Dublin, dead since 1866), " The Marine Algæ of North America," in Smithsonian Contributions, 1858, three volumes, with fifty plates.

The chief authority on North American fossil plants is Leo Lesquereux, a native of Switzerland, residing in Columbus, Ohio, who is besides a trustworthy judge of mosses, and compiled the catalogue of Arkansas plants in Owen's Geological Report. An important branch of science, the geographical distribution of American plants, is yet in its infancy. It requires a thorough knowledge of local floras in connection with the physical and climatological condition of each locality to get the right view of this matter. Some steps have been made in that direction, some preparatory work has been done, but the main labor is left to the future. Prof. Gray published "Statistics of the Flora of the Northern U. S.," in *American Journal of Science and Arts*, 1856, which will promote the cause for that part of the country. Dr. J. G. Cooper published a good article on the distribution of the forests and trees of North America, in Smithsonian Report for 1858. As only the woody plants are here accounted for, the limits drawn cannot be intended as to separate botanical districts in general. Even for the forest plants the limitations admit of some corrections, but as a preliminary essay it is valuable.[1]

Here this sketch must be concluded, for two reasons—1, the newest botanical literature is so extensive, and partly published in so many different periodicals, that a private library is not sufficient for a survey of the whole; 2, the number of botanists has increased so much throughout the country, as is shown by Cassino's Naturalist's Directory, that it is rather difficult to winnow the chaff from the wheat, and to avoid offence by neglecting a man whose merits are worthy of mention.

———:o:———

[1] Although published in 1860, yet, as the author has mentioned several works published later than 1850, we may here draw attention to the Flora of the Southern United States, containing abridged descriptions of the Flowering Plants and Ferns of Tennessee, North and South Carolina, Georgia, Alabama, Mississippi and Florida, by A. W. Chapman, M. D. The Ferns, by Daniel C. Eaton, New York, 1860. pp. 621.—EDITORS.

PART II

SELECTED PAPERS
FOR DIFFERENT REGIONS

INTRODUCTION

John W. Harshberger

INTRODUCTION.

Philadelphia lies in a nearly level plain, on the western bank of the River Delaware, in 39° 57′ 7.5″ N. latitude, and 75° 9′ 23.4″ west from Greenwich. The city is 86 miles from the Atlantic Ocean by the Delaware River, 125 miles in a direct line north-east of Washington, and 85 miles south-west from New York.

It is situated in a rich agricultural region, protected from the sweeping western and north-western storms by the range of hills known as the Blue Ridge. When first settled by white men, the region lying within 60 miles radius of the city, including New Jersey, was densely wooded with a great variety of fine forest trees, which, growing upon rich agricultural soil in south-eastern Pennsylvania, were rapidly cut down with the spread of cultivation. This region was the favorite haunt of the Delaware Indians. Intersected by two great streams, the Delaware and Schuylkill Rivers, any part of it could be reached by hunting parties in a short time by water. Into these two rivers, numerous creeks and rivulets run, swelling the volume of water which empties into the ocean at Capes May and Henlopen, and supporting a variety of important food-fishes, such as the salmon, shad, trout and cat-fish. Under cover of the trees and watered by the numerous streams which intersect the country, a surprisingly large number of herbaceous plants is to be found, which, together with the rich variety of graceful forest trees, give a peculiar charm to the entire district. In early days, the scenery must have been impressively beautiful

before the marring hand of man disturbed the equilibrium
of nature.   Forest and plain, streams and rivers tumbling
over numerous cascades, rocky, fern-clad ravines, high hill
summits give, even at the present day, a diversity to the
landscape.   Two or three spots, preserved in their primitive
naturalness, still attest to the wild attractiveness of the
scenery, which, nowhere very bold or grand, gives to the
country a peculiarly peaceful aspect, in harmony with the
moods of the early Quaker settlers.   Two such places still
preserve the quiet beauty of the early river scenery, namely:
the Wissahickon and the Brandywine regions, a stream of
the former name emptying into the Schuylkill in Fairmount Park, and one of the latter name into the Delaware
near Wilmington.   The Wissahickon is one of the most
romantic of American streams.   The slope of the land on
each side is high and abrupt.   Self-guarded by these rock
battlements, it retains a primeval character.   Along its
banks trees and vines hang down to the water's edge, and
numerous springs drip from the rocks.   Its unbroken quiet,
its dense woodland, its pine-crowned hills, its sunless
recesses and sense of separation from the outer world contrast strongly with the broad meadows, flowing river, and
bright sunshine of the adjacent region.

The topography of the district is no less marked than
the general landscape.   To the east of the Delaware, the
low-lying plain of southern New Jersey, with an elevation
at a few points of from 200 to 300 feet above sea level, is a
very striking feature.   This plain geologically dates its
origin to the cretaceous and tertiary periods, and is made
up of alluvium along the Delaware River and Atlantic
Ocean beaches, and of yellow gravel, glass sand and sandy

RAPIDS, WISSAHICKON CREEK, FAIRMOUNT PARK.

clays, composing by far the greater extent of the so-called West Jersey tertiary formation, with the exception of a narrow band of the cretaceous green sand and marl beds, potter's clay, fire sands and clay, which parallel its course with the Delaware River, extending in a north-east direction to Raritan Bay. The western water-shed is traversed by streams, which, rising in the marl district and yellow-gravel region of the interior, flow into the Delaware, being affected in their lower reaches by tide-water. The eastern water-shed is intersected by several important streams, such as Mullica, Great Egg Harbor and Toms Rivers. These rivers mainly take their rise in cedar swamps and sphagnum bogs for which the region is noted. North of the marls, as we approach the mountians, a region in which red shale mainly predominates, is entered upon. West of the river, an undulating plain along the river front rises gradually to the older paleozoic hills, which reach an elevation of two hundred feet or more. Back of these, as the Blue Ridge is approached, the country becomes more undulating and broken by numerous hills of various geological formations.

Enough has been said by way of introduction to show that these topographic, hydrographic and geologic features have an important bearing on soil formation, and consequently on plant life and distribution. We find that each topographic, hydrographic and geologic district has some plants peculiar to it. Each of the plant communities, into which the flora of a district as large as Philadelphia can be divided, can be distinguished by the component plants, which, together with their collective features, give character to the vegetation of the particular geological, topographical

or hydrographical region. Such a flora as that of Philadelphia, comprising in New Jersey and Pennsylvania some
1200 species at the outside, can be classified into several ecological communities, such as the Hydrophytic, Halophytic
and Mesophytic, the first of which, by way of example,
may again be further sub-divided into those societies which
comprise the water plants growing in the Delaware and
tributary streams and Atlantic Ocean, such as the Plankton
Society, the Hydrocharite Society, the Nereid Society, the
Sea Grass Society, Schizophytic Society, Reedy Swamp
Society, the Swamp Society, the Sphagnum Bog Society,
the Cedar Swamp Society, etc.

The peculiar attractiveness of the region and the richness of the flora have so enticed botanists into the field that
systematic botany has been almost exclusively the depart-·
ment of the science practiced by a majority of those mentioned in this work.   Then, too, a living was not to be
had by the prosecution of botany in America in the early
days. It was pursued solely as a pastime and a healthy
recreation by busy men, physicians, bankers and merchants.
We find, however, in looking over the list of names, that
wherever botany was pursued as the main object of life,
that those men, who thus devoted their entire time to the
science, became famous. Excluding names of the present
generation, John Bartram, Humphrey Marshall, Zaccheus
Collins, William Darlington, Elias Durand, John Evans, A.
P. Garber, Joshua Hoopes, Peter Kalm, Adam Kuhn, James
Logan, Isaac Martindale, André Michaux, G. H. E. Muhlen-
berg, Lewis D. von·Schweinitz, Thomas Nuttall, W. P. C.
Barton, Charles Pickering, Frederick Pursh, C. S. Rafinesque,
John Redfield, and David Townsend, achieved distinction

VIEW OF WISSAHICKON CREEK, FAIRMOUNT PARK.

along systematic lines. It was not until after the perfecting
of the microscope and the epoch-making period, beginning
with issue of Darwin's Origin of Species, that the modern
study of botany may be said to have begun in Philadelphia. The pursuance of botany in Philadelphia and in
America generally can be divided into four periods:

(1) The early descriptions of the flora by persons not con-
versant with botany, who described the plants after the manner of the old herbalists, chiefly as interesting rarities, or as
useful, natural medicines. The sect of German Pietists
presided over by Kelpius, established in 1694 on the lower
Wissahickon, a garden where medicinal plants were raised
for use and study. It may, therefore, be styled the first
garden in America where a botannical arrangement of
plants was made.* In 1739 was published at Leyden,
in Holland, an essay in Latin, entitled, " Experimenta et
Meletemata de Plantarum generatione," by the learned
Governor of Pennsylvania, James Logan. It was afterwards, in 1747, republished in London, with an English
translation, by Dr. John Fothergill. The experiments and
observations were admirably illustrative of the doctrine of
sexes of plants † established by Jacob Camerarius. This
may be said to be the first work of any botanical importance issued by a Philadelphia botanist. Many of Logan's
ideas smack of medieval scolasticism, so that he is properly
placed in the Pre-Linnæan period.

(2) The period of the ascendency of Linnæan ideas.
John Bartram was one of the first persons who may be said

---

*SACHSE. *The German Pietists of Pennsylvania*, p. 75.
† See an article of mine, " James Logan," *Botanical Gazette*, Aug., 1894.
1889. SACH'S *History of Botany*, 391-392.
1849. DARLINGTON—*Memorials of Bartram & Marshall*, 21.

to have used the Linnæan system in the study of plants. Dr. Benjamin Franklin introduced Bartram to European botanists, among them Doctor Gronovius, who presented the Quaker botanist with Linnæus's Systema Naturæ of 1740.* The overwhelming influence of the great Linnæus gave to the botany of the eighteenth century an almost exclusively systematic and descriptive character. Linnæus was the author of the binomial system of nomenclature of plants and animals, which still goes back to his work as its basis, and of the artificial "sexual system" of classification based on the stamens and pistils of the flowering plants, whose functions, as reproductive organs, were already realized. The order which he brought out of the chaos of descriptive natural history was a blessing so unalloyed, and his system was so simple and seductive, that it was many years before most botanists again began to realize that their science properly comprehends other problems than those involved in naming and pigeon-holing plants.

It was while the Linnæan enthusiasm was at its height that the first Philadelphia botanists appeared on the scene.

In the year 1748, Peter Kalm, a Swedish naturalist, and pupil of Linnæus, visited Pennsylvania and spent three years in exploring America, and in 1753 published his travels.† Doctor Adam Kuhn, of Philadelphia, was proba-

---

*1740. Linnæus—*Systema naturæ, in quo naturæ regna tria, secundum classes, ordines, genera, species systematice proponuntur Editio II auctior. Stock- holmiæ, Gottfr. Kiesewetter.*

Bartram's copy of this book is in possession of the Pennsylvania Historical Society; on the title page is the writing: "John Bartram His booke sent to him by Dr. Gronovius in ye year 1746."

That it is authentic is shown by the following, also written in the book: "I bought this book June 14, 1853, at the sale at Mackey's of Books of Col. Carr, who married Bartram's grand-daughter." E. D. Ingraham. "I bought this book March 20, 1855, at the sale of Mr. Ingraham's Library by M. Thomas & Sons." A. Day.

†1753-61. P. Kalm—*En Resa til Norra America.* Stockholm, III vols.
1754-64. Kalm—*Beschreibung der Reise nach dem nördlichen Amerika.* Göttingen. 3 Theile (German translation).

DEVIL'S POOL, WISSAHICKON CREEK
(circa 1885).

bly the first professor of botany in America, appointed in 1768 to the chair of botany in the University of Pennsylvania. He had the advantage of studying under the illustrious Swede, and was said to have been a favorite pupil (Linnæo ex discipulis acceptissimus). John Bartram next becomes pre-eminent as a botanist. In the latter end of the year 1785, Humphrey Marshall published his Arbustum Americanum,* a description of the trees and shrubs native of the United States. It is the first strictly American botanical work. In 1791 William Bartram's Travels † appeared, and in 1801 André Michaux's ‡ "Oaks of North America." Two years later, in 1803, the first elementary work on botany by Prof. B. S. Barton, § was published in Philadelphia.

F. André Michaux, ‖ in 1810, issued his splendid history of the Forest Trees of North America (Histoire des Arbres Forestiers de l'Amérique Septentrionale) with elegantly colored plates. An excellent catalogue of the native and naturalized plants of North America was published by Dr. Henry Muhlenberg at Lancaster, in 1813.¶ Later, Frederick

*1785. HUMPHREY MARSHALL—*Arbustum Americanum, the American grove or an alphabetical catalogue of forest trees and shurbs, natives of the American United States.* Philadelphia.

† 1791. WILLIAM BARTRAM—*Travels through North and South Carolina, Georgia, East and West Florida, etc., containing an account of the soil and natural productions of those regions.* Philadelphia.

‡ 1801. ANDRE MICHAUX—*Historie des chênes de l'Amérique, ou descriptions et figures de toutes les espèces et variétés de chênes de l'Amérique septrionale.* Paris (folio).

§ 1803. B. S. BARTON—*Elements of Botany; or outlines of the natural history of vegetables.* Illustrated by forty plates. Philadelphia.

‖ 1810. FRANCOIS ANDRE MICHAUX—*Histoire des arbres forestiers de l'Amérique septentrionale, considérées principalement sous les rapports de leur emploi dans les arts et de leur introduction dans le commerce, ainsi que d'apres les avantages, qu'ils peuvent offrir aux gouvernements en Europe, et aux personnes, qui veulent former de grandes plantations.* Paris.

¶ 1813. MUHLENBERG—*Catalogus Plantarum Americæ Septentrionalis huc usque Cognitarum, Indigenarum et Cicurum;* or, a Catalogue of the Hitherto Known Native and Naturalized Plants of North America. Arranged according to the Sexual System of Linnæus. Lancaster, 1813. Wm. Hamilton, octavo, pp. iv., 112.

Pursh published in London, in 1814, his valuable and comprehensive work, Flora Americæ Septentrionalis.*

Arranged according to the Linnæan system there appeared in 1818, in two volumes, Dr. William P. C. Barton's†
Compendium Floræ Philadelphicæ, a hastily digested, but thoroughly useful hand-book of the region.

Botanical works and papers began now to multiply, and the third period of Philadelphia botany was fairly entered upon with the publication in 1818 of Nuttall's "Genera of North American Plants," at Philadelphia.‡

(3) Development of the Natural System under the influence of the doctrine of the constancy of species. A new direction to the study of systematic botany, and morphology was given in France, where the sexual system had never met with great acceptance. Bernard de Jussieu and his nephew, Antoine Laurent de Jussieu, taking up Linnæus' profounder and properly scientific efforts, made the working out of the natural system, in Linnæus' own opinion the highest aim of botany, the task of their lives. The key was given by the study of the order Ranunculaceæ in the Jardin des Plantes. In 1789 Jussieu's System appeared. It was not until 1815 that the natural system of Jussieu was received by the botanists of Philadelphia. In that year Abbé Correa published for the use of his class in Philadelphia a reduction of the genera of Muhlenberg's Catalogue according to the system of Jussieu. This was

---

*1814. PURSH—*Flora Americæ septentrionalis, or a systematic arrangement and description of the plants of North America.* London, II vols.

†1818. W. P. C. BARTON—*Compendium Floræ Philadelphicæ, containing a description of the indigenous and naturalized plants found within a circuit of ten miles around Philadelphia.* Philadelphia, II vols. 8. I. Preface 251 pp. II. 234 pp. cum indices.

‡1818. NUTTALL—*The Genera of North American Plants, and a catalogue of the species of the year 1817.* Philadelphia, II vols.

appended to a second edition of the catalogue issued in 1818 by Solomon Conrad, and was probably the first attempt in the United States to group our plants by the natural method.

In 1826, in conjunction with some of his intimate friends, Dr. William Darlington, of West Chester, assisted in organizing the Chester County Cabinet of Natural Science, of which institution he was president from its origin; in the same year he published his "Florula Cestrica,"* being a catalogue of plants growing around the borough of West Chester, Pennsylvania. This paved the way for a large and more comprehensive manual of the botany of Chester County, which appeared in 1837 under title of "Flora Cestrica."† A third edition of this book appeared in 1853. This work at the time of its issue was one of the most complete local floras extant, and is still a model for all works of a similar character. The descriptions are clear, lucid and minute, and its use even to-day is not replaced by a manual of more modern issue.

The study of the cryptogams received a great impetus at the hands of Lewis D. von Schweinitz, who published in 1831 a synopsis of North American fungi, "Synopsis Fungorum in America Borealia Media Digentium."‡

Elias Durand, one of the most acute systematists of his

---

*1826. DARLINGTON—*Florula Cestrica : an essay towards a catalogue of the phænogamous plants, native and naturalized, growing in the vicinity of the borough of West Chester, in Chester County, Pennsylvania, with brief notices of their properties and uses in medicine, rural economy and the arts.* West Chester, 4 min., pp. xv., 152, 3 tab. col.

†1837. DARLINGTON—*Flora Cestrica : an attempt to enumerate and describe the flowering and filicoid plants of Chester County, in the State of Pennsylvania.* West Chester, 8. xxiii, 640 pp. 1 map col.

‡1831. SCHWEINITZ—*Synopsis Fungorum in America Borealia Media Digentium.* Trans. Amer. Philos. Soc. N. S., IV p. 141 (177 pp., 4to., 1 plate).

day, who, if he had had proper encouragement, would have been one of the shining lights in the botanical firmament, contributed several botanical papers to the *Journal of the Academy of Natural Sciences*, namely, descriptions of Heermann's and of Pratten's collections.*

The views of European botanists were undergoing a change under the influence of the history of development and knowledge of the minuter anatomy and embryology of the cryptogams (1840–1860). Schleiden's " Grundzüge der wissenschaftlichen Botanik " † appeared, but its chief title is Die Botanik als inductive Wissenschaft, which indicates the point on which Schleiden laid most stress. His great object was to place the study, which had been so disfigured in the text-books, on the same footing with physics and chemistry, in which the spirit of genuine inductive enquiry into nature had already asserted itself in opposition to the nature-philosophy of the immediately preceding years. This change in European thought does not seem to have had much effect on the botanists of Philadelphia, who were busy in working up the plants collected in various parts of North America, both by private individuals and by the botanists of the trans-continental surveys.

(4) The year 1860 may be said to mark the beginning of the modern era of botany. Darwin's Origin of Species,‡

---

*PLANTÆ HEERMANNIANÆ—*Descriptions of New Plants collected in South California, by Dr. A. T. Heermann,* Naturalist attached to the Survey of the Pacific Railroad route, under Lieut. R. S. Williamson, by E. Durand and Theo. C. Hilgard. Journ. Acad. Nat. Sci. 2nd ser., III, 37–46.

†1842–43. SCHLEIDEN—*Grundzüge der wissenschaftlichen Botanik, nebst einer methodologischen Einleitung als Anleitung zum Studium der Pflanze.* Leipzig. 2 Theile.

1845–46—Second edition. (Die Botanik, als inductive Wissenschaft behandelt.)

‡1859. DARWIN—*On the origin of species by means of natural selection ; or, the preservation of favored races in the struggle for life.* London. John Murray octavo pp. ix., 502.

published in 1859, was an epoch-making book. It introduced the modern period of scientific thought.

With the exception of Thomas Meehan, Joseph T. Rothrock, Thomas C. Porter, Charles Pickering, John H. Redfield, Thomas P. James, Benjamin M. Everhart, Rev. Francis
Wolle, Mary Treat, William P. Wilson, J. Gibbons Hunt,
Emily L. Gregory, John M. Macfarlane, Job B. Ellis, George
Rex, H. C. Wood, Henry Trimble, Edson S. Bastin, Ida
Keller, Henry Kraemer, J. W. Harshberger and H. C. Porter,
very few of the Philadelphia botanists have advanced materially the science of botany according to the progress made
in morphology, physiology and taxonomy. The others
have unfortunately given their attention to herborizing,
and have overlooked the deeper and more interesting problems which are still to be worked out, such as the reasons
underlying the geographical distribution of the plants in
the region, phenological inquiries or the philosophy of
the time of flowering; physiological problems suggested
by growth and development, and ecological questions suggested by the environmental conditions. It is to be hoped,
however, that with the modern training to be had at
several institutions of learning, our botanists will give up
discussing the differences between species already described
and will devote their energies to advancing modern
botanical thought. The facilities for those who desire to
obtain a modern botanical training are many. The oldest
botanical centre, namely, the University of Pennsylvania,
presents, in its Biological School, a place where such
instruction may be had.

A history of the development of botany in connection

with the University of Pennsylvania is interesting.* "So far as now appears, Dr. Adam Kuhn, a pupil of Linnæus, was the first botanical professor in Philadelphia; or in the country, being appointed in the year 1768. There is, however, no record of any important work connected with his name. As early as the year 1800, Dr. Benjamin Smith Barton was teaching botany in Philadelphia, and numbered among his pupils in 1803–'04, at the University of Pennsylvania, William Darlington, who subsequently became known as one of the most learned and exact botanists of his day in this or any other country. Dr. Darlington says of his preceptor, 'that he did more than any of his contemporaries in diffusing a taste for the natural sciences among the young men who then resorted to that school.' He also published in 1803 'the first American elementary work on botany, at Philadelphia.'"

"The minutes of a trustee meeting held April 7, 1812, show that 'a letter was received from Dr. Barton requesting the use of one of the rooms in the University to deliver his lectures on natural history and botany in.' The request could not be granted. In July, 1813, Dr. Barton resigned his professorship of materia medica, a position which does not appear to have been a bed of roses. He was succeeded by Dr. Chapman. The following minute appears of a trustee meeting of November 7, 1815:"

"*Whereas*, the Legislature of Pennsylvania, by their Act passed the 19th March, 1805, granted to the trustees of this institution out of the moneys due to the State, the sum of three thousand dollars, for the purpose of enabling them

* I have drawn largely at this point on Dr. J. T. Rothrock's sketch of the Biological School, published in the Circular of Information Bureau of Education, entitled, "Benjamin Franklin and the University of Pennsylvania" (1893).

SYSTEMATIC BEDS AND INSTITUTE, UNIVERSITY BOTANIC GARDEN
(LOOKING WEST IN 1896).

to establish a garden for the improvement of the science of botany, *Resolved*, that Mr. Rawle, Mr. Chew and Mr. Burd be a committee to consider and report the best method of carrying the said intention of the Legislature into effect."

"February 6, 1816, at a trustee meeting Mr. C. S. Rafinesque and Dr. William P. C. Barton offered themselves as candidates for the professorship of natural history and botany in the University. Dr. Barton was appointed."

"The trustees received March 19, 1816, 'a letter from a society of gentlemen called the Cabinet of Sciences, relating to a botanical garden. It was referred to the committee on that subject. Mr. Binney and Mr. Gibson were added to the committee on botanical garden.' On April 2, the committee was authorized to solicit subscriptions from the public towards the accomplishment of that end. Nothing having been accomplished by meeting with the Cabinet of Sciences, on April 16 the committee announced that they had published their application for aid in the public papers. By order of the board, the moneys available for the botanical garden were put at interest, subject to future call. Early in 1817 forty-two acres of ground had been purchased for the botanical garden. The records show that it was located in Penn Township, near the 'Canal Road,' and it was ordered that enough for the purposes of the garden should be 'fenced off.' "

"Stringent economy had apparently become a necessity, and in 1819, after two years' ownership, the trustees were considering the propriety of selling the ground purchased for a botanical garden, and the professor of botany was 'allowed the use of the yard south of the University, as the same is now inclosed, for the cultivation of plants there, at his own expense, during the pleasure of the board.' "

" On October 4, 1818, the faculty of natural history was instituted, and the following professorships created: First, botany and horticulture; second, natural history, including geology, zoology, and comparative anatomy; third, mineralogy, and chemistry, as applied to agriculture and the arts."

"The only signs of life in 1820 in the department of science were now the appointment of a committee to consider the propriety and the cost of erecting a greenhouse, and the request from the janitor that he be allowed the use of Prof. Cooper's room for the winter, to preserve the plants ' he had collected to adorn the grounds and to encourage the love of botany.' The request was granted. The report of the committee on the greenhouse was laid on the table."

Prof. Barton, in 1822, writes to the board that he had lectured in the winters of 1816, 1817, 1818, 1819, 1820, 1821, and further, that he had refused to receive the fees from the students. The botanical instruction in 1821 was discontinued because a class could not be formed. The crisis in the school of natural history, however, was reached in March, 1827. It appears that no lectures had been given for several years by the professor of natural history, including geology, or by the professor of comparative anatomy, and that the professor of botany was then holding the professorship of materia medica in the newly-started Jefferson Medical College. Early in 1828 the faculty of natural history was abolished.

"Now, however, it appears that the medical faculty, which would have no botany while Dr. Barton occupied the chair, had become suddenly solicitous about that science, and, as a result, the trustees re-established the chair of botany in 1829, placing it on the same footing as it was

UNIVERSITY BOTANIC GARDEN (LOOKING TOWARD BIOLOGICAL SCHOOL).

before the institution of the faculty of natural science, and Mr. Solomon W. Conrad was speedily chosen to fill it. The appointment was probably the best that could have been made." Mr. Conrad, who died in 1831, was, as stated by one of his contemporaries, an "amiable man," and an "excellent botanist," was probably the earliest to "attempt to group our plants by the natural method."

Dr. George B. Wood was elected to the chair of materia medica in the University in 1835. In addition to the creation of an admirable cabinet of drawings and specimens illustrative of materia medica, Dr. Wood erected a spacious greenhouse, in connection with a garden, and stocked them with many varieties of rare tropical and exotic plants, which he exhibited as illustrations of the subjects treated in his lectures. In 1865 Dr. Wood endowed an auxiliary faculty of medicine in the University of Pennsylvania, including a chair of botany, to which his nephew, Dr. Horatio C. Wood, was appointed in 1866. He held this professorship for ten years, resigning the chair of botany for that of materia medica and therapeutics, made vacant by the death of Prof. Joseph Carson. Dr. Joseph T. Rothrock was elected to fill the vacancy caused by the removal of Dr. H. C. Wood to the chair of materia medica and therapeutics, a position which he still holds. Botany, under his direction, received a great stimulus, when on December 4, 1884, the School of Biology, erected by the liberality of Dr. Horace Jayne, was opened to students. Teaching began at once, with modern biological methods. Later Dr. William P. Wilson was appointed Professor of the Anatomy and Physiology of Plants, in conjunction with Dr. Rothrock, who devoted himself to the systematic side of botany. All of the departments of botany, since the establishment of the school, have received consideration at the University. Morphology, taxonomy, physiology, paleobotany, economic botany, forestry, pathological and geographical botany, have been taught at various times ; chief stress, however, being laid on morphology, taxonomy and physiology, as the departments of botany most necessary to students. A post-graduate class in botany, composed of student candidates for the degree of doctor of philosophy, has been maintained. The teaching force of late years, consisting of Drs. Rothrock, Wilson, Macfarlane, Harshberger and Porter, has maintained the standard desirable in a modern school of botany.

The Herbarium of the University, through the generosity of Mr. Isaac Burk, possesses a singularly complete representation of the flora of the vicinity of Philadelphia, consisting of about six thousand specimens from this and other localities. Mr. Aubrey H. Smith presented by will his excellent herbarium, which, with the collection made by the late Joseph Leidy, forms a most excellent working herbarium. Many specimens from the earlier government expeditions, and suites of the collections made by Parry, Hall, Barbour, Vasey, Bolander, Palmer, Lemmon, Canby, Ward, Pringle, Bebb, Wolfe, Curtis, Reverchon, Rothrock, Harshberger and others, are represented. The herbarium also contains a large proportion of our native ferns, mosses and lichens, and over two thousand species of fungi, all of which have been carefully determined. A museum of economic botany was started by Dr. Rothrock in connection with the School of Biology, and further additions were made in material collected on his cruise to the West Indies in the winter of 1889-1890.

POND, UNIVERSITY BOTANIC GARDEN (LOOKING SOUTH).

The University Botanic Garden was begun with the erection of the building for the School of Biology. It consisted, in 1888, of about a quarter of an acre of ground immediately surrounding the Biological School, planted with a few systematic and experimental beds. The planted grounds were surrounded by high gravel banks, overgrown with weeds. It was not until 1890, when a large part of this glacial gravel deposit had been sold and carted away, that the botanic garden may be said to have had its inception. Dr. Joseph T. Rothrock, Professor of Botany, supervised the laying out of the ground to the east and west of the laboratory, which was planted to grass, with trees and shrubbery arranged for landscape effect. A tank pond of considerable size was also built for the growing of various water-plants. A lean-to conservatory for the growth of hot-house plants was also a feature of the garden at this time. The ground, as laid out by Dr. Rothrock with systematic regard to the position of the plants, included finally about an acre of ground surrounding the laboratory building. Several rare shrubs were set out, among them, *Neviusia Alabamensis,* an anomalous rosaceous plant found growing wild in the Southern states. The grass plots, shrubbery and systematic beds then occupied a terraced depression fronting on Pine Street.

The development of this garden, however, took place when Mr. C. C. Harrison accepted the provostship of the University. In 1893, immediately after his appointment to be Professor of Botany, Dr. John M. Macfarlane submitted plans for the establishment of the botanical garden, on the triangular piece of land back of the biological laboratory. Various circumstances conspired to prevent the carrying out of these plans until the autumn of 1894, when Dr. Macfarlane was asked to become Professor-in-Charge of the Biological School. Through the fostering care of Provost Harrison and Vice-Provost Fullerton, the work steadily advanced under the direction of Prof. Macfarlane. The gravel bank, overgrown with weeds, rapidly assumed its present pleasing appearance.

There are over 3000 distinct specimens growing in the gardens, while nearly 1500 more are all but ready for planting. The lawns are 300 feet in length, the eastern lawn being 157 feet long and 110 feet wide, subdivided into 44 small beds, whose dimensions are 45 feet in length by $3\frac{1}{2}$ feet in width. The western lawn is an almost exact counterpart of the eastern lawn. The beds contain a large number of species of plants, arranged systematically according to the Engler and Prantl system. The plants are arranged and labeled with the scientific and common name, the native place or habitat and the medicinal property, if any. The donations of seeds and plants to the garden include gifts from the botanical gardens of Edinburgh, St. Petersburgh, Dublin, Jena, Cambridge and other European botanical centres. On the terraced area further back a physiological grouping of flowering plants is now being made. Here separate beds are given to climbing, tendril-bearing, succulent, spiny, insectivorous, variegated and other series. Thus similar changes produced by environment on species that have no systematic affinity can be graphically demonstated to the student.

The arboretum is from three to five acres in extent, and will only be excelled by those of Harvard University and the Shaw Gardens, near St. Louis. The greater part of the

UNIVERSITY BOTANIC GARDEN IN 1898 (LOOKING NORTH).

property will be devoted to the arboretum, which already contains a number of trees of interest and beauty. These are planted in systematic order along the drive-way which enters on Woodland Avenue and encircles the garden. A magnificent, ornamental bed, fashioned somewhat after the beautiful bed in the famous Kew Gardens, in London, is a feature. It is 200 feet long and 8 feet wide, and is filled with herbaceous plants. Unlike the series of small beds before referred to, it will not be a scientific feature, but will be the chief ornament of the gardens. The plants are so arranged as to present a succession of flowers from early spring to late fall.

The contractor in excavating left a deep cut in which the pond, bog garden, iris bed, rockery and fernery are situated. The pond, of irregular shape, this last season (1898) was filled with a splendid growth of aquatics, water lilies, lotuses and water hyacinths being conspicuous, while the aquatics, *Marsilia quadrifolia, Myriophyllum, Nitella, Chara Limnobium, Limnocharis* and *Trianea bogotensis* grew luxuriantly. The bog garden is situated along the ditch connecting the Victoria tank and the lake. In separate pockets formed by stones set on end are grown plants which flourish in a water-logged soil, such as *Decodon verticillatus, Acorus calamus, Typha latifolia, Sparganium eurycarpum, Drosera rotundifolia, Sarracenia purpurea, S. flava, Helonias bullata, Orontium aquaticum*, species of *Carex*, of *Cyperus*, of *Sagittaria*, of *Juncus*, and a host of others too numerous to mention. The iris bed adjoins the bog garden, and is connected with it by a pipe through which a water supply is furnished to the roots of the plants. The rock garden covers the sides of the cut in which the lake is situated, and is provided with separate pockets for every plant after the rockery in the botanic garden at Edinburgh. Here are grown a large number of rock plants and herbaceous ones of a gaudy color. Narrow pathways intersect the rock garden in every direction, so that a person can study the plants closely, as well as in mass. The fernery, hardly yet thoroughly established, is in a glen through which runs a cindered path under trellis-work devoted to climbing plants, intended to protect the delicate ferns beneath. Nearby is the Bryarium for the growth of mosses.

The surrounding shrubberies have been laid out so as to illustrate geographic groupings of plants. One is devoted to the swamp shrubs of the eastern States, such as the white azalea, white birch, spice brush, swamp magnolia, andromedas, huckleberries, cedar and juniper. Another includes the rhododendrons, azaleas and kalmias of our woods. Under the shade of these, native and introduced herbaceous plants thrive, that would soon shrivel if exposed to hot suns.

Through the generosity of Provost Harrison important additions were made to the plant houses at the close of the season of '97. These houses now represent more than 9000 feet of glass surface, and consist of eight houses in addition to propagating frames. One of the greenhouses, immediately connected with the laboratory for plant physiology, is in part utilized as a temperate house, in part for the work of students in plant physiology. An adjoining house, 34 × 11 feet, is arranged as a fernery, and contains a representative collection of ferns and their allies. Opening from the last are a propagating house, 40 × 10 feet, a stove house, 46 × 18 feet, and a palm house, 59 × 28 feet. The two last now

VIEW IN THE PALM HOUSE, UNIVERSITY BOTANIC GARDEN.

contain a varied collection of pitcher plants, aroids, melasto-
mids, sensitive plants, palms, marantas, bananas, bamboos,
etc. To the right of the palm house is a succulent house
containing a type collection of cacti, euphorbias, gasterias,
aloes, agaves, crassulas and other forms that are more or
less similarly modified to live in arid regions and successfully resist long periods of drought. On the left side of the
palm house are two structures, each 59 × 13 feet. The
inner of the two now contains a fair collection of sub-
tropical and tropical orchids donated by Mr. LeBoutillier,
and more recently by Mrs. George Wilson. Sharing the
house with these are parent species and hybrid derivatives
of the popular begonias and gloxinias, as well as specimens
of the curious South African genus *Streptocarpus*, two species of which show only one of the two seed leaves, though
this may attain a length—as in one specimen exhibited in
the greenhouses—of two to three feet. Species of *Oxalis* and
Solanum, the curious simple-leaved *Chorizema* from Australia,
and many other sub-tropical types of great value in under-
graduate and graduate teaching find a home here. The
outer or cool house lodges many plants of great botanical
interest, chief among these being the celebrated venus fly-
trap, several native sundews, groups of our southern sarracenias, and the butterworts, all celebrated as fly catchers.
Recently, by permission of the highway authorities of the
city of Philadelphia through a municipal act, Pine Street,
between Thirty-eighth and Thirty-sixth Streets, has been
taken from the city plans. The area thus vacated has been
converted (1898) into a fine walk lined with trees, shrubs
and rhododendrons. At the Thirty-ninth Street entrance a
memorial gate-way, in keeping with the dormitory building

adjoining, has been erected by the Class of '73. A vivarium or building for small animals is in course of erection in the garden enclosure immediately in the rear and to the west of Biological Hall. A small garden is much better for scientific work than a large one, the cost of maintenance of the latter being considerable. The University garden of five or six acres is therefore admirably adapted to its purpose, being near to the laboratory where the botanical instruction is given. The illustrations will convey better than words the appearence of the garden after it had been planted in 1896, and again after the construction of the greenhouse additions and vivarium in 1899.

The Botanical Society of Pennsylvania was instituted at the University of Pennsylvania, October 23, 1897. Under its auspices a fortnightly series of popular meetings and of scientific meetings have been held since organization, while during the fall, spring and summer, courses of laboratory demonstrations and field excursions have been held. A great variety of interesting papers were presented during the first year of the society's existence. Living plants from various greenhouses, charts, diagrams, lantern slides and specimens added very much to the attractiveness of the several meetings. The class meetings were held at the Biological Hall of the University, where the greenhouses and garden afforded much interesting and valuable material. The general meetings were held in the auditorium of the Harrison Chemical Laboratory. The following persons have interested themselves in the movement : Dr. John M. Macfarlane, Professor of Botany; Dr. Henry Kraemer, Messrs. Roberts LeBoutellier, W. H. Walmsley, Drs. A. W. Miller,

BOG AND ROCKERY, UNIVERSITY BOTANIC GARDEN
(LOOKING NORTH IN 1896).

H. C. Porter and J. W. Harshberger. A list of the active members of the society is given in an appendix.

The Philadelphia College of Pharmacy has also been an influential botanical centre. Several excellent botanists have occupied the chair of materia medica and botany, as John M. Maisch, Edson S. Bastin, Henry Kraemer and Clement B. Lowe. The chemical and pharmaceutical side of botany have been much emphasized, and much meritorious work has been done, both by the chemists and botanists of the institution. *The American Journal of Pharmacy* is a valuable epitome of the work accomplished.

The late Professors Trimble and Bastin, of the faculty, were actively engaged in botanical research, the former on the tannins of plants, the latter on the coniferæ and the resins. From the College of Pharmacy many students have received an inspiration for botanical study. The Herbarium of the Philadelphia College of Pharmacy possesses the collections of Elias Durand, Daniel B. Smith, Prof. John M. Maisch, and that of Isaac Martindale, purchased by Messrs. Smith, Kline, French and Company from the estate, as also numerous contributions from botanical friends and students. With a laboratory equipped for botanical and microscopical study, and with such an excellent herbarium for comparison, the College is enabled to give an extended course in botany.

The Academy of Natural Sciences of Philadelphia was founded March 21, 1812, by a few citizens "interested in the study of the works and laws of the Creator." From the outset, the Department of Botany received a due share of attention, and the first contribution to the Academy's Herbarium * consisted of a collection of plants made in the

---

* Torrey Bulletin VIII: 42, J. H. Redfield.

environs of Paris and presented by Nicholas S. Parmentier, and still in excellent preservation. During the years which followed, this little nucleus received constant accretions from the working botanists of the day, and the names of Collins, Elliott, Pursh, Baldwin, LeConte, Conrad, Nuttall, Torrey and Pickering are inscribed on many of the early tickets of the Herbarium.

In 1834 the Academy received the bequest of the collections made by Rev. Lewis David von Schweinitz during a period of forty years. Most of the northern species were collected by himself, but many came from Dr. Torrey, Major Le Conte, Rev. Mr. Dencke, and other correspondents. The European species were contributed by Weldon, Bentham, Brongniart, Schwaegrichen, Steudel and Zeyher. The Siberian plants were furnished by Ledebour, and those of India by Wallich and Steinhauer. Many Chinese species were collected by Mr. James Read, and from the Arctic regions were plants collected by the navigator Parry, and received through Sir William Hooker; while from South America were rich collections made by von Martius, Huffel, Hering and Baldwin. Perhaps the most valuable portion of the bequest consisted of the extensive series of the lower cryptogams, of which Von Schweinitz had made a special study.

Other valuable contributions followed the bequest of Von Schweinitz, among which may be specified the Poiteau collection of St. Domingo plants; Chilian plants from Dr. Styles and Dr. Ruschenberger; Nuttall's collections made in his expeditions to Arkansas, Oregon and the Sandwich Islands; Menke's Herbarium of 7000 species of European plants derived from Thunberg, Sprengel, Bernhardi,

GROUP OF SARRACENIAS BY THE POND, UNIVERSITY BOTANIC GARDEN.

Treviranus, Mertens, etc.; the Ashmead collection of marine algæ; Lesquereux's collection of over 700 species of algæ, authenticated by the best algologists of the age, and a large collection of cryptogams from Ravenel. More recent additions are the herbaria of the late Thomas G. Lea, of Cincinnati, and of Dr. Joseph Carson, late Professor of Materia Medica in the University of Pennsylvania; a large collection from southern Europe and from India, made by the late John Stuart Mill, received from Miss Taylor, through the Director of the Kew gardens and the kindness of Dr. Gray; the collections of the late Dr. Charles Pickering, made in his journeys through oriental regions in 1844 and 1845; Syrian and Algerian plants from Dr. George Post, of Beirut; Floridan plants from Dr. Garber; Mexican plants collected by Parry, Palmer, and Pringle, and a set of mosses and hepaticae of North America, collected and named by the late Col. F. Austin.

The most important accession to the Academy's collection was the Short Herbarium of Dr. Charles W. Short, of Louisville, Ky. For this the Academy was indebted to the strenuous exertions of Dr. Gray in its behalf, and to the liberality of Dr. Short's family. The plants of this collection are uncommonly choice specimens, from all active collectors up to 1863, and are laid in sheets of extra size, arranged in 325 book-form cases, of which the North American species occupy 261, and the exotic species 64.

The work of arranging the earlier collections of the Academy was mainly accomplished by Nuttall and Pickering, followed later by Goddard, Bridges, Zantzinger, Durand, Burk, Scribner, Redfield, Smith, Brown and Meehan. Until the removal to the new building, in 1876, the arrangement had been after the Linnæan system in large cumbersome port-folios, in a narrow, dark and inconvenient hall. The removal gave opportunity for an entirely new arrangement, more in accordance with the progress of the science, on enclosed shelves after the most approved modern methods, and in well-lighted apartments convenient for reference and study.

In 1854, the lamented Elias Durand began the work of forming a special North American Herbarium from the stores of the Academy, contributing largely from his private collection, of species collected by Lindheimer, Fendler, Wright and others. In this labor he was occupied four years. Since his death the work of perfecting this department has been continued, and nearly all of the collections made in our newer territories by Parry, Lemmon, Palmer, Kellogg Ward, Rothrock, Pringle and others have been contributed at various times by Gray, Canby, Parker, Meehan, Rothrock, Martindale and Redfield. This collection and the " Short Herbarium " occupy the upper of the two rooms devoted to botany in the south-west corner of the building, while the lower room contains the general herbarium, and a large case devoted to the reception of fruits, seed vessels and other vegetable productions.

One of the most recent additions to the Academy's Herbarium is the loan collection of the Lewis & Clark plants from the American Philosophical Society. The following is an interesting account of this recent acquisiton :

" The expedition of Captains Merewether Lewis and William Clark, from what was then the village of St. Louis to the sources of the Mississippi and across to the Pacific Coast, was one of the marvels in the early history of the American

DRIVE-WAY, UNIVERSITY BOTANIC GARDEN (LOOKING NORTH).

Republic.* Captain Lewis started from Washington to take charge of the party on the 5th of July, 1803. They crossed the Continent, reaching the mouth of the Columbia River, and with the loss of but one man, returned and arrived at St. Louis on the 23d of September, 1806.

"The idea of exploration originated with Jefferson. In 1792 he tried to interest the American Philosophical Society in the plan. It was approved, and it was decided to place the expedition in charge of André Michaux. Reasons of State policy arising out of our relation with Michaux's country, caused its abandonment. Lewis was Jefferson's private secretary, and under him the expedition finally started."

The plants collected on the expedition were described by Pursh in his " Flora Americæ Septrionalis," published in London, in 1814. One hundred and nineteen (119) plants are referred to, many of which he described as wholly new.

Nothing was known as to the final disposition of the collections. It was lost to botanists. " It was understood that Pursh took these plants to England, and that they were left by him to Mr. A. B. Lambert, Vice-President of the Linnæan Society, under whose roof and by whose aid Pursh's great work was completed. Lambert's Herbarium was finally distributed, and, in some way not known to the writer, a number of Lewis's plants, forming Pursh's types, and marked 'from Lambert's Herbarium' became part of the herbarium of the Academy of Natural Sciences of Philadelphia."

" Two years ago Professor C. S. Sargent suggested to the writer the possibility of some of the material being yet in

*1898. MᶜEHAN—Proc. Acad. Nat. Sci., p. 12.

the custody of the American Philosophical Society. After long and diligent search, packages of plants were found which could only be these, as the localities on the label slips were about the same as those given in Pursh's work." After a careful scrutiny of the labels, handwriting and plant sheets it was satisfactorily determined by Mr. Meehan that the plants were those of Lewis and Clark. Pursh had evidently studied these collections before starting to Europe with them, leaving duplicates, where there were any, and those which were too imperfect to be easily recognized. A comparison of Lewis's own labels and Pursh's copies shows that the latter were not always strictly copied—differences can be seen in the comparisons made in the catalogue. Pursh's notes were probably made from Lewis's original memoranda carried away with the specimens, and are, therefore, the more likely to be the exact statements of the collectors, than the copies left with these. The plants first determined by Mr. Meehan were turned over to the Gray Herbarium where they were critically studied by Messrs. Robinson and Greenman.

With the freedom of three-quarters of a century the museum beetles had made sad work in the bundles. In a few cases the specimens had been wholly reduced to dust, and only fragments were left in other cases. Generally, however, they were in fair condition. The Philosophical Society wisely accepted a proposition to deposit these and other collections with the Academy of Natural Sciences, where they would be properly cared for. All these collections, including those from the Kuram Valley, Afghanstan, made by Major J. E. T. Aitcheson; from China, Japan, Formosa, Australia and Tasmania; from the Texo-Mexican region ; from Australia, made by Baron F. von Mueller; from the United States Forestry Commission of rare North American trees; from North Africa, made by Geo. Curling Joad; from the North Pacific Survey, by William Canby; from Alaska, by Thos. Meehan; from the Yellowstone, made by F. Tweedy ; of Mexican plants distributed by C. G. Pringle, the noted collector, and the veteran botanist, Dr. Palmer; from Colorado, New Mexico and California, made by A. H. Smith; from Chili, Bolivia and Brazil, distributed by H. H. Rusby; from Tabasco and Chiapas, in Mexico, by Prof. Rovirosa; from South America, by Thos. Morong; from the West Indies, made by Professor Leopold Krug, of the Royal Botanical Museum, Berlin; from Guatemala, distributed by John Donnell Smith; from Greenland, made by Wm. E. Meehan; from Greece, Macedonia, Asia Minor, Kurdistan and Mesopotamia, by Bornmüller and Sintenis; from the West Indies, distributed by Rothrock ; from California, by Brandegee, are valuable scientifically, because they represent type specimens of the new forms discovered by all of these collectors in different parts of the world. In addition to the phanerogams the Academy's herbarium has been enriched in recent years by the addition of many noteworthy cryptogamic collections, among these may be mentioned a complete set of Ellis's "Centuries of North American Fungi," Drummond's "Mosses of the Rocky Mountains and British America," a set of fungi, from the wife of the late Dr. Geo. Martin, of West Chester; the lichen herbarium of Dr. J. W. Eckfeldt, the celebrated lichenologist, and other collections of minor interest and importance.

With these large collections the herbarium of the Academy of Natural Sciences may be said to be on a par

with those of Harvard University, at Cambridge, Mass.;
Columbia College, in New York; the Missouri Botanical
Garden, at St. Louis, and the United States Department of
Agriculture, at Washington.

In addition to the herbarium, the Academy is especially
fortunate in having an almost complete file of all of the
leading journals of science, in which list the botanical
journals are well represented. The Academy, therefore, is
well equipped for active scientific work, but is hampered,
like so many other institutions, by lack of funds. It is to
be hoped that the endeavor which is now being made to
raise an endowment to pay a first-class botanist, and to
maintain the herbarium in good condition, will meet with
success. The fund, to be known as the Redfield Memorial
Herbarium Fund, is sorely needed, as the committee, consisting of Thomas Meehan, George M. Beringer, and Stewardson
Brown, testify in their appeal to the admirers of the scientist
who did so much for the herbarium.

It is estimated that at least $30,000 should be raised
to insure the necessary income, and the bequest * of Mr.
Redfield will serve as a nucleus. It is proposed to utilize
the interest to pay a conservator or professor, who shall
devote his time to the needs of the herbarium, and make
the collections available at all times. Any income in
excess of the sum needed for salary will be judiciously
applied to shares in exploring expeditions, or other means
of adding to the collections.

The Philadelphia Botanical Club, organized by Dr.
J. Bernard Brinton, who held the presidency until his death,
has for its object the promotion of social intercourse between

* See Science N. S. I : 470; also *Philadelphia Ledger*, April 2, 1895.

botanists who live within a radius of sixty miles of the City of Philadelphia, the formation of a herbarium in which all of the plants of the region, carefully mounted, labeled and annotated, are represented, and the advancement of botany generally. Field trips during the spring, summer, and autumn months are taken to various points of botanical interest, and reports are made at each succeeding meeting of the plants collected. Its membership represents the active botanists of the region at the present day. It has done much to advance the systematic knowledge of the plants of the district. Under the auspices, and with the co-operation of this organization, Dr. Ida Keller has undertaken the preparation of a list of the plants found within the neighborhood of Philadelphia, as represented in a radius of 60 miles or less. This work will be of great use to students of the local flora, and is to be highly commended.

The study of the lower forms of plant-life has been almost entirely neglected by the greater number of botanists mentioned in this work. In order to create an interest in the fungi, especially the higher fleshy fungi, two societies have been inaugurated during 1897 and 1898. One called the Philadelphia Mycological Center, modeled after the Boston organization, meets statedly at the Academy of Natural Sciences. Topics of general interest to the members are discussed, and specimens, chiefly of the edible kinds of toadstools, are presented for inspection.

The other organization is known as the Mycological Club. Its objects are essentially similar to those of the first-mentioned society. A bulletin is published under the auspices of this club, and excursions are taken into the surrounding country for specimens. Those interested especially in the advancement of the interests of this club are the following ladies and gentlemen: Captain Charles McIlvaine, Mrs. S. T. Rorer, Dr. Henry Leffman, Mr. and Mrs. Talcott Williams, and Theodore Rand, C. S. Ridgway, Dr. S. C. Schmucker, and Mrs. Theodore Ely.

The Pennsylvania Forestry Association, organized in 1886, has done a great service to the State of Pennsylvania in interesting its people in trees and in forest preservation. As an outcome of this agitation under the leadership of the Forest Commissioner, Dr. J. T. Rothrock, and by the official organ of the Society, " *Forest Leaves*," three tracts of mountain land have been designated as forest reservations.

The Delaware County Institute of Science at Media, Penna., founded in 1833, has for its object the diffusion of general and scientific knowledge among its members and in the community at large, and the establishment and maintenance of a library and historical record and a museum. The library of the Institute contains about four thousand volumes, covering generally the subjects of science, history and literature. The museum contains a large collection of specimens, illustrating the fauna, flora and the minerals of Delaware County. The local botanical and mineralogical collections are quite complete, well arranged, and accessible to students of these subjects. The Indian archeology of the county is well represented. The Institute is divided into several sections, as follows: biological anthropological, physical and literary sections.

The Wagner Free Institute of Science, at 17th and Montgomery Avenue, Philadelphia, was founded by William Wagner to advance the cause of science by popular lectures

MAIN HALL, HORTICULTURAL BUILDING, FAIRMOUNT PARK.

and demonstrations. In the past many lectures on botany have been given to interested audiences, under the auspices of the Institute, which also possesses a fine scientific and general library.

Fairmount Park and its Horticultural Building also are places where the botanists of Philadelphia have received their inspiration. This building, in Moorish style, was built for the Centennial Exposition of 1876, and in it was placed a large and valuable collection of palms, orchids, tree ferns, ferns and other tropical and exotic plants. It has been altered considerably since it was built to. give more light to the rapidly-growing araucarias, palms and bamboos. A visitor luxuriates in the vegetation of the fernery, the forcing-house, the temperate-house and the main hall, in which grow some magnificent specimens of Australian palm (*Ptychosperma elegans*) tree ferns, bamboos, traveler's tree, date palms, rubber trees, fan palms, climbing aroids, wax palms, and other tropical plants. Upon entering the door, one imagines himself in a tropical forest.

The Commercial and Economic Museum,* which is owned and operated by the City of Philadelphia, is composed of the combined exhibits of many countries, both of raw material and the vegetable and animal products of the countries represented. This museum was established soon after the close of the Columbian Exposition.

Professor Thomas Meehan and Professor W. P. Wilson, made the proposition to procure these great collections to one or two public-spirited gentlemen on September 7, 1893, and on September 12th of that year a resolution was

---

* *The Mirror*, Philadelphia, Wednesday, May 8, 1895, with portraits of those prominently connected with the Philadelphia Museums.

passed by the Select and Common Councils of the City of Philadelphia authorizing the Park Commission to make collections for an Economic Museum. Later, arrangements were made by Professor Wilson and one member of the Park Commissioners with the Mayor, by which letters were addressed to the foreign representatives at the Columbian Exposition, stating the wish of the City of Philadelphia to obtain the exhibits of natural products at the Fair for the proposed Museum, where they might be preserved intact and so remain as a lasting proof of the advancement of the countries they represent.

On October 19, 1893, Councils passed an ordinance making an appropriation of $10,000 to the Commissioners of Fairmount Park "to defray the expenses of procurement, transportation, packing, storing and display of raw and manufactured economic products now of the Columbian Exposition at Chicago." The sum of $3000 was advanced by three prominent citizens until such time as Councils should make the appropriation. This unexpected generosity saved the enterprise from what might have been a failure, since by that time other cities and institutions, realizing the benefit to their industries to be gained by such a museum, were making attempts to obtain the collections partially promised to Pennsylvania. The appropriation was finally made by Councils, and the money was judiciously expended, $20,000 provided for the project in 1894. Professor Wilson succeeded in securing displays of various materials from Mexico, from Costa Rica, from Guatemala, from British Guiana, from Ecuador, from Colombia, from the Argentine Republic, Brazil, Venezuela, Uruguay, Paraguay, Labrador, Sweden, Germany, Russia, Johore, Japan, Siam, New South

FERN HOUSE, HORTICULTURAL BUILDING, FAIRMOUNT PARK.

Wales, Turkey, British India, Persia, Spain, Puerto Rico and Ceylon.

The objects of the museum are clearly set forth by its promoters as being: First, to bring before American manufacturers all the varied products of the world, that they may make the best selection of material for their own especial interests. Second, to publish all possible scientific and useful information concerning these products which may aid the manufacturer and consumer in his choice. Third, to place on exhibition manufactured articles and samples, with full information from all markets which ought to be entered or controlled, and to furnish to merchants and manufacturers useful information concerning opportunities in foreign lands.

The exhibits consist in the main of raw materials, showing the vegetable and animal products of the several countries, as for instance the handsome forestry exhibit from Mexico, composed of a great number of prepared woods, many of them polished and varnished on one side, showing the grain and any particularly striking features of the wood. There are also minor forest products, such as fibres, gums, resins, tannins and medicinal plants. In many cases the collections represent big sums of money, the exhibition from the Argentine Republic, alone, having cost that government over $25,000. One of the three collections presented from Japan cost $15,000 to prepare. The collections from many of the countries are of especial interest to botanists, in that they comprise largely a display of the vegetal productions of those lands.

The Museum, being in need of a building sufficiently large to accommodate the vast quantity of material in its possession, there were assigned nineteen rooms in City Hall,

all of them except three being in the basement. Many cases were stored in the warehouses of several firms in Philadelphia, awaiting a time when they might be opened.

The exhibits continued in the City Building until September, 1895, when they were taken to South Fourth Street, a lease of the Pennsylvania Railroad Company's Buildings, which are admirably adapted to the purposes of the museum, having been made with the Pennsylvania Railroad Company at advantageous terms in August of that year. The buildings now occupied have been leased for five years, and the exhibits will remain in them until the buildings are completed in West Philadelphia. These railroad buildings are three in number. The principal one is the granite building, fronting on Fourth Street at the corner of Willing's Alley. Adjoining it also on Fourth Street is the Empire Building, three stories in height. Connected with the granite building is the rear of the annex, an enormous structure six stories high. Altogether, the museum occupies 128 rooms with a floor space of 200,000 square feet.

Part of the granite building is devoted to the display of exhibits according to products, without regard to the geographical location of the countries producing them. Here are shown samples from every civilized section of the world, embracing everything of foreign growth used or deemed capable of being used by American manufacturers, or which enter into or are likely to enter into American commerce. The exhibits include thousands of samples of woods, wools, silks, cottons, vegetable fibres, hides, skins, dye-stuffs, tanning materials, drugs, herbs, minerals, coffees, spices, teas, rubber, etc.*

---

* *Philadelphia Inquirer*, Monday, March 2, 1896.

Another section of the Empire Building is given over to the American forestry exhibit, particular attention being given to the Southern states, which are just now being looked to in a commercial sense as they have never been before. A large part of this display was secured at the Atlanta Exposition, and includes the collection of sugar cane from Louisiana, and the interesting turpentine exhibit, showing realistically the method of collecting this valuable product of the turpentine forests.

The exhibits are tastefully and conveniently arranged according to countries, beginning with Mexico and following with the Central and South American countries, in their order. After these come the countries of Europe, Asia and Africa.

Especial prominence is given to Mexico and the Spanish-American countries because of their growing importance to the mercantile and manufacturing interests of this country. This prominence, however, is not at the expense of the exhibits from other countries, for the collections from all of them will be extensively and conveniently displayed. The exhibit from Mexico can be taken as an illustration of the completeness of the different collections. It occupies no less than .nine large rooms, and embraces every possible article of commercial value that country produces. In the exhibit are collections of woods from no less than fifteen different states in the Mexican Republic, which have already been or will be in the near future brought into use by the manufacturers of this country.

Another department, which illustrates the great scope of the museum and the thoroughness contemplated in its general plan, is the testing department. Here,. with suitable machinery and under the supervision of experts, will be made tests, for instance, of samples of foreign woods for the purpose of ascertaining their availability for certain uses. The scientific laboratories of botany and zoölogy and those of technology in connection with the museums are doing excellent work in the study of economic samples.

A department, fully as invaluable to the American manufacturer as any of the others, is that in which are displayed samples of foreign manufacturers. This display consists of a complete collection of manufactured articles which certain countries, notably those of Spanish America, Australia, South Africa, etc., do not produce themselves and which they must necessarily purchase elsewhere. An inspection of this department will show an American manufacturer just what these countries buy and where they buy.

A Bureau of Information is maintained whose object is to make a special study of foreign commerce, compile all data relative thereto, and make it available to the manufacturer or consumer in as concise and definite a form as possible. The bureau is located on the third floor, and a force of men and women is actively engaged in compiling the data, arranging indexes and getting things in shape.*

A library† is maintained in connection with the Bureau of Information, where business directories, trade and commercial publications, books of reference, etc., from all parts of the world are kept constantly on file. The library is receiving between 400 and 500 of the best trade publications from England, France, Germany and the United States, over fifty of them coming from London alone.

---

* See *Ledger*, February 19, 1896.
† *Philadelphia Inquirer*, March 2, 1896.

In addition to these are the official organs of Great Britain, France, Germany, Russia, Italy, Australia, Japan, Mexico and the South American countries.

There is also kept a complete file of statistical documents issued by different countries in relation to trade and commerce. The information and data contained in all of these publications is compiled and indexed for ready reference under the most approved library methods, so that the merchant or manufacturer may easily and quickly find that which refers to the particular line of industry in which he is interested.

The authorities expect, in the near future, to move the collections to West Philadelphia, near the University of Pennsylvania. On June 27, 1895, City Councils passed an ordinance giving over to the Trustees eight acres of land along the Schuylkill. By an ordinance approved October 10, 1896, eight acres more were added to this, making sixteen acres. Recently $200,000 has been appropriated out of the "loan bill" to commence work on the buildings; $50,000 was granted by the State of Pennsylvania; $100,-000 has been raised by private subscription; and in December, 1898, the Congress of the United States passed a bill, which was signed by the President, authorizing the expenditure of $350,000 in the erection of exhibition and museum buildings for the Philadelphia Commercial Museums, so that the museums have become a national as well as a state and municipal enterprise.*

A casual reader will see, after perusing this sketch of

* Since writing the above, exposition buildings have been started and are well under way. An Exposition and Commercial Congress, it is planned, will be held in Philadelphia, beginning with the middle of September, 1899. It is planned that two of the exposition buildings, under course of erection, will become a permanent part of the Commercial Museums.

the facilities which are presented at Philadelphia, that the city is peculiarly fitted to be the botanical centre of America. Situated between New York, the metropolis of America, and the Capitol of the United States, it is within easy reach of the metropolitan life and publishing houses of the former city, and the libraries and scientific departments of the latter city, in the Smithsonian Institute and National Museum, and in the National Congressional Library. The libraries of the Pennsylvania Horticultural Society, the American Philosophical Society, the Pennsylvania Historical Society, the University of Pennsylvania, the Franklin Institute, the Free Library Company, and the Philadelphia Library Company present unusual opportunities for research and study. In addition to the facilities for study and research already mentioned, the city has Fairmount and Bartram's Parks, and the seed houses of national reputation of Landreth, Dreer, Buist, Blanc and Burpee, whose experiment farms lie within close proximity to the urban limits. Philadelphia has never very severe winters, being protected by the range of hills to the west and north-west. Lying in close proximity to New Jersey, whose peculiar flora is rich in species, and to the drainage areas of the Susquehanna, Delaware and Schuylkill Rivers, it is favorably situated for botanical research. Why not make Philadelphia the Botanical Centre of America?

BOTANY AND BOTANISTS
IN NEW ENGLAND

James Ellis Humphrey

BOTANY AND BOTANISTS IN NEW ENGLAND.

By James Ellis Humphrey.

IT was natural that the earliest students of plants were interested in them chiefly as sources of sustenance or of healing. It was wholly from this point of view that they were treated by the classic writers, whose accounts furnished the only available information for the few who interested themselves in such matters during the Dark Ages. And long after the revival of learning they remained very few who were not deterred by the fate of Roger Bacon and Albertus Magnus from the investigation of natural phenomena. When finally the results of study and travel had shown the futility of the early attempts to identify the Mediterranean plants of Theophrastus and Dioscorides with those of northern Europe, attention was turned from study of the books to the examination of plants. But it was still chiefly for their real or reputed medicinal value that they were thought of interest, and the *materia medica* came shortly to include almost every known plant. To very many of these the most marvellous properties were attributed by the herbalists and the gatherers of simples.

When New England was settled, there was no science of botany. The only knowledge of plants was this pseudo-science of the drug-mongers, and the most elementary principles of their activity in the economy of nature were yet undreamed-of. Theology held full sway over men's minds, and knowledge of the physical world suffered equally under both prevailing forms. The dominant theology of the greater part of Europe often served as a cloak and excuse for an idle and a brainless life. The opposing Calvinism which ruled New England saw in our world only a "vale of tears," and staked its existence on keeping it so; while its logic was to glorify the Creator by contempt for creation. Among the early writers on New England, the Winthrops, Dudley, Higginson and Wood made incidental reference to some plants of the country, and the earlier explorers had carried back a few of the more striking ones to Europe. Thus by the beginning of the seventeenth century our Indian corn, pitcher-plant, milk-weed and arbor-vitæ, among others, were known to the herbalists.

The first serious attempt to describe the natural products of our region was that of an Englishman, John Josselyn, who spent several years here. The chief authority for English-speaking students of his time were the herbal of Gerarde, published in 1636, and that of Parkinson, which appeared four years later. These are huge quartos, filled with the most fantastic statements about plants and their virtues, and illustrated by crude but, for the most part, recognizable wood-cuts of the plants described. The accompanying illustration is from a photograph of Parkinson's herbal. The page on the left is devoted to the peanut and may give an idea of the so-called botanical knowledge of the times.

This volume of 1755 pages is called the *Theatrum Botanicum*, and was prepared by "John Parkinson, Apothecary of London, and the King's Herbarist." At least so far as American plants are concerned, it justified its claim to contain accounts of more "than hath been hitherto published by any before." The author's classification well shows the utter lack of insight into plant structure and the emphasis laid on their properties in his time. Some of his groups are:

"1. Sweete smelling Plants. 2. Purging Plants. 3. Venomous, Sleepy, and Hurtfull Plants, and their Counterpoysons. 4. Saxifrages or Breake-stone Plants. 5. Vulnerary or Wound Herbes."

In a few cases, where the plants of what we now call a natural group or family possess some striking external peculiarity, this was seized upon; so that a few of his groups correspond in a general way with some now recognized. Such are:

"8. Umbelliferous Plants. 11. Pulses. 13. Grasses, Rushes, and Reedes."

This was the state of knowledge when Josselyn wrote; and, as he was an amateur and no very accurate observer at best, we shall not find his work superior to that of his contemporaries. He made visits to New England in 1638 and in 1663, the latter of several years' duration; and most of the time was spent as the guest of his brother, a settler at Black Point, now Scarborough, Maine. He published two volumes concerning these travels, one entitled, "New England's Rarities discovered in birds, beasts, fishes, serpents and plants of that Country," and published at London in 1672; the other, "An Account of two voyages to New England," the second edition of which appeared in 1675. The latter has been reprinted in book form in Boston, and also in the "Collections" of the Massachusetts Historical Society. The former was issued with full and valuable annotations by Prof. Edward Tuckerman, in 1865; and careful notes on that part of it relating to plants had previously been published by Rev. J. L. Russell, in *Hovey's Magazine of Horticulture* for 1858. Both works contain extensive notes on the products of the region, quite in the style of the times. The lists given in the "Rarities," especially "of such plants as have sprung up since the English Planted and Kept Cattle in New England," have been of service in throwing light on questions concerning the indigenous or introduced character of certain plants, and concerning the time of arrival of various vegetable immigrants. It will afford an instructive idea of the botanical and therapeutic notions of our earliest American ancestors to quote Josselyn's accounts of two well-known plants. He gives the earliest figure and description of our jewel-weed, *Impatiens fulva*.

"This Plant the Humming Bird feedeth upon, it groweth likewise in wet grounds, and is not at its full growth till July, and then it is two cubits high and better, the Leaves are thin, and of a pale green Colour, some of them as big as a Nettle Leaf, it spreads into many Branches, knotty at the setting on, and of a purple Colour, and garnished at the top with many hollow dangling flowers of a bright yellow Colour, speckled with a deeper yellow as it were shadowed, the Stalks are as hollow as a kix, and so are the roots, which are transparent, very tender, and full of a yellowish juice. . . . The Indians make use of it for Aches, being bruised between two stones, and laid to cold, but made, (after the *English* manner) into an unguent with Hogs Grease, there is not a more sovereign remedy for bruises of what kind soever; and for Aches upon Stroakes."

The description is surprisingly good, and the figure unmistakable. A hundred years later the same plant was thought an excellent remedy for jaundice. To-day it has no place in the *materia medica*.

Of America's best-known gift to the world Josselyn says:—

" The vertues of Tobacco are these, it helps digestion, the Gout, the Tooth-ach, prevents infection by scents, it heats the cold, and cools them that sweat, spent spirits restoreth, purgeth the stomach, killeth nits and lice; the juice of the green leaf healeth green wounds, although poysoned; the Syrup for many diseases, the smoak for the Phthisick, cough of the lungs, distillations of Rheume and all diseases of a cold and moist cause, good for all bodies cold and moist taken upon an emptie stomach, taken upon a full stomach it precipitates digestion, immoderately taken it dryeth the body, enflameth the bloud, hurteth the brain, weakens the eyes and the sinews."

A suggestion of the current mythology comes in the query appended to the account of plants introduced from England:

" 2. What became of the influence of those Planets that produce and govern these Plants before this time? "

And the strictly human standpoint from which all diseases were regarded could hardly be better shown than by the following:—

" Their fruit-trees are subject to two diseases, the *Meazels*, which is when they are burned and scorched with the Sun, and lowsiness, when the woodpeckers job holes in their bark: the way to cure them when they are lowsie is to bore a hole into the main root with an Augur, and pour in a quantity of Brandie or Rhum, and then stop it up with a pin made of the same tree."

However popular with human patients this treatment may have been, our author had hardly learned its efficacy for fruit-trees by personal experience.

Before the first native botanist is encountered, more than a century must be passed over. But this period was a great one in the history of science. The latter part of the seventeenth century saw the founding of the Royal Society of London and of other learned societies devoted to the "increase of natural knowledge," which were the fruits of the great awakening to the external world brought about by the Baconian philosophy. In spite of the fact that natural science speedily became the vogue in England, and a chief subject of polite conversation, some of its greatest discoveries date from this time. Such are those of the law of gravitation and of the circulation of the blood. Plants now began to be philosophically studied, and the broader outlines of a classification based on really fundamental features were now sketched. Besides, improvements in the theory and construction of the microscope made possible its application to the study of the minute structure of plants, and many important discoveries concerning their physiology were made in these and the following years. But the overwhelming influence of the great Linnæus obscured these profounder studies, and gave to the botany of the eighteenth century an almost exclusively systematic and descriptive character. Linnæus was the author of the binomial system of nomenclature of plants and animals, which still goes back to his work as its basis, and of the artificial "sexual system" of classification, based on the stamens and pistils of the flowering plants, whose functions as reproductive organs were already recognized. The order which he brought out of the chaos of descriptive natural history was a blessing so unalloyed, and his system was so simple and seductive, that it was many years before most botanists again began to realize that their science properly comprehends other problems than those involved in naming and pigeon-holing plants.

It was while the Linnæan enthusiasm was at its height that the first New England botanist appeared on the scene. Very many American plants had already been described by Linnæus and his followers; but they had been collected chiefly by travellers or settlers from Europe, and few of these had visited our rocky corner of the country. Manasseh Cutler was a native of Killingly, Conn., and was graduated at Yale in 1765. After a few years of teaching in Dedham, Mass., and of business life at Edgartown, in the intervals of which he read law, he was admitted to the bar. But he soon decided to enter the ministry, and studied with his father-in-law, Rev. Mr. Balch, of Dedham. He received calls from several Massachusetts parishes, and finally, in 1771, accepted that to Ipswich Hamlet, which became, in 1793, the town of Hamilton. He there remained until his death, in 1823, the peer of any amongst the greathearted and large-minded ministers of those "times that tried men's souls," in whom New England orthodoxy so far outran in practice the meagre promise of its theory. His energy and capacity enabled him to study and successfully practise medicine during the Revolution, when the parish physician was at the front, and to eke out his slender salary by fitting for college many boys, in the list of whom one finds not a few of Salem's distinguished names,—Lowell, Silsbee, Derby and Cabot among the rest,—and by teaching the theory of navigation to others who became the most famous ship-masters of that old port's palmy days. After the Revolution, o'erleaping the narrow confines of New England, he personally engineered through Congress the famous grant of land in the Northwest Territory on which the first settlements were made, and secured the incorporation of the momentous clause in the "Ordinance of 1787" excluding slavery from the territory. He gave two sons to the first settlement at Marietta, and soon after spent a few months there, making the first serious study of the age of the wonderful earth-works of the Ohio valley, from data furnished by the trees and remains of trees found on them. He served his district in Congress during Jefferson's first term.

It was while doing double duty as spiritual and physical healer that Dr. Cutler first became attracted to the study of the plants of his neighborhood. He had also been much interested in some problems of plant physiology by reading the still classic "Vegetable Staticks" of Stephen Hales; but the difficulties under which he labored may be gathered from the fact that not even a barometer could be had in Salem. The dearth of books and of money for their purchase in a country parish, and the practical isolation occasioned by a separation of twenty-five miles in the days of horse-power locomotion, were also hindrances of a very practical kind to the pursuit of science.

In 1781 Dr. Cutler wrote to the Corporation of Harvard College that he had been trying to study plants, but had not the necessary books and had failed to procure them in Europe. He therefore asked to be allowed to borrow "Dr. Hill's Natural History" from the College library for a short time, offering to pay for its use such sum as the Corporation might determine. In spite of his modesty, one may be quite sure he received the desired permission unconditionally. In the same year, when the American Academy of Arts and Sciences was organized, Dr. Cutler was elected a member, and, two years later, a member of its first committee on communications. In 1785, in the first volume of the Memoirs of the Academy, was printed the chief published result of this writer's botanical studies, and the first account, after Josselyn's, of New England plants. This paper is entitled, "An Account of some of the Vegetable Productions growing naturally in this part of America, botanically arranged." It shows the beginning of the scientific spirit in botany, but also shows how large a place was still given by students of plants to the investigation of their properties. But the discrimination now evident is in marked contrast with the wholesale credulity of the previous century.

In his introduction Cutler wrote:

"The almost total neglect of botanical enquiries, in this part of the country, may be imputed, in part, to this, *that Botany has never been taught in any of our Colleges*, and to the difficulties that are supposed to attend it; but principally to the mistaken opinion of its inutility in common life."

A few extracts may illustrate his treatment of common plants:—

"HAMAMELIS.— *Witch Hazel.* . . .

"The Indians considered this tree as a valuable article in their *materia medica*. They applied the bark, which is sedative and disentient, to painful tumors and external inflammations. A cataplasm of the inner rind of the bark is found to be efficacious in removing painful inflammations of the eyes. . . . The specific qualities of this tree seem, by no means, to be accurately ascertained. It is, probably, possessed of very valuable properties."

Whatever may be its real value, witch-hazel has not yet lost its popular reputation for the very virtues here ascribed to it.

"ASCLEPIAS. — *Silkweed.* . . .

"It may be carded and spun into an even thread, which makes an excellent wick-yarn. The candles will burn equally free, and afford a clearer light than those made of cotton wicks. They will not require so frequent snuffing, and the smoke of the snuff is less offensive."

Here is perhaps a practical suggestion for these days of the revival of candle-light.

"BERBERIS.— *Barberry.* . . .

"It is said that rye and wheat will be injured by this shrub, at a distance of three or four hundred yards; but only when it is in blossom, by means of the *farina fecundans* being blown upon the grain, which prevents the ears from filling."

This is an interesting statement of the explanation offered a hundred years ago of the harmful influence of barberry bushes upon grain, the belief

in which dates back to a much earlier time. It is now known that the microscopic dust wafted to the grain from the barberry bushes is not, as Cutler thought, the pollen of the latter, but the spores of a parasitic fungus, which passes through a part of its life-history upon the barberry, and requires to be transferred to a grain or other grass to complete the cycle.

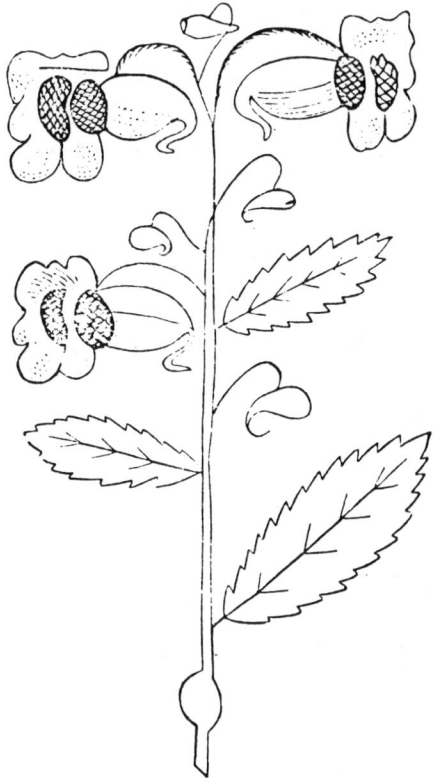

JOSSELYN'S "HUMMING-BIRD PLANT."
FROM "NEW ENGLAND'S RARITIES DISCOVERED."

In July, 1784, Dr. Cutler was a member of the first party to ascend the White Mountains for scientific observation; and he repeated the trip just twenty years after. An outcome of this was the "List of trees and plants," which he furnished for Belknap's History of New Hampshire. On all his journeys he studied plants unweariedly. He had doubtless projected an elaborate botanical work, and the manuscript materials he left show that this would have placed him in the front of American botanists. More than a dozen volumes of notes and drawings were accumulated by him, and some of them are still in existence. According to the late Professor Tuckerman, he anticipated much of the work of later botanists, such as the separation of the hickories from the true walnuts, and the indication of many of the new species first published by Bigelow, Nuttall and Gray.

He conducted a voluminous correspondence with the most distinguished men of America and with many in Europe. The complete set of the "Species Plantarum" of Linnæus, sent him by the Swedish botanist Swartz, is now in the library of the Essex Institute, at Salem. He was elected a member of many learned societies, and received the title LL. D. from his *alma mater* in 1789. He was a lover of plants from an æsthetic, as well as from a scientific, standpoint, and cultivated about his house many not before seen in New England, such as the pawpaw, the persimmon, the tulip-tree and the trumpet-vine. In the variety of his interests and occupations, and in his fine grasp of whatever he undertook, our pioneer botanist is but another example of the efficacy of hard work; and his life of eighty-one years is additional evidence that, with reasonable care, the human mechanism wears out no sooner than it rusts out.

An attempt to supply the lack of botanical teaching in our colleges, which Cutler remarked, was soon thereafter made. The first instruction in natural history in New England was given in lectures by Dr. Benjamin Waterhouse, "Professor of the Theory and Practice of Physic in the University at Cambridge, Mass." A native of Newport, R. I., and trained in the best European schools, Dr. Waterhouse steadily maintained that his interest in natural history was wholly subsidiary to his devotion to

medicine; but his conception of the scope of botany was so broad and modern, and his lectures were so superior, with due allowance for the state of knowledge in his time, to much that passes for botanical teaching to-day, that his standpoint deserves recognition here. He held for a few years the chair of Natural History in Brown University, and gave a course of twelve lectures on that general subject in 1786–87. In 1788 he began annual courses of lectures at Harvard, which attracted much attention. The botanical lectures of these courses were printed in the "Monthly Anthology" from 1804 to 1808; and in 1811 they were published at Boston in a volume entitled "The Botanist." In the preface to this volume, the author says:—

"It is of importance that one universal language should be adopted by botanists; but it is wrong to make that and classification the primary object. Agreeably to this doctrine is the sentiment of the famous Rousseau, who, in his *Letters on the Elements of Botany*, says, 'I have always thought it possible to be a very great botanist without knowing so much as one plant by name.' . . . To be able to pronounce, at first sight, the name of each mineral, to distinguish one genus of plant from another, and to discriminate stuffed animals in a museum were, it seems, enough to entitle a man to be considered a Natural Historian; when, at the same time, he perhaps knew nothing of the anatomy of a seed, and of its gradual development into a perfect plant and flower, producing again a seed or epitome of its parent, capable of generating its kind forever."

These words are equally appropriate at present, for the idea of botany which they embody is hardly better grasped by the popular mind than when they were written.

A result of Dr. Waterhouse's lectures was that "several gentlemen of opulence and literary influence in the government of the University came to the resolution of laying a foundation for a Professorship of Botany and Entomology; to which they determined to annex an extensive Botanic Garden." Thirty or forty thousand dollars were subscribed, and the Legislature gave two townships toward the

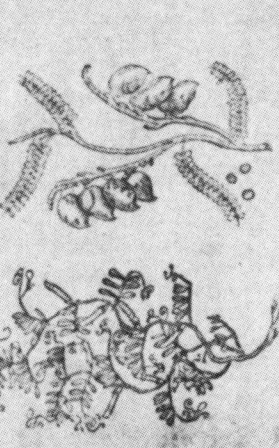
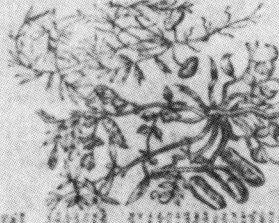

PARKINSON'S THEATRUM BOTANICUM, 1640.

MANASSEH CUTLER.

establishment of the professorship and garden at Cambridge. The Massachusetts Society for promoting Agriculture also aided the garden project.

In 1805 the present Botanic Garden was established; and William Dandridge Peck became the first incumbent of the chair of Natural History, which he held for seventeen years. A student and an ingenious mechanic, Professor Peck was especially interested in insects, and his influence upon the botany of his time was not marked. After his death Harvard was too poor for twenty years to fill his chair; but Thomas Nuttall, a distinguished naturalist of English birth, was called to the curatorship of the Botanic Garden, and remained in that post until 1828. A printer by trade, Nuttall's love of travel and of natural history had led him a roving life. Except for his six years at Cambridge, his headquarters during thirty-six years residence in America were in Philadelphia. During this time he travelled in all parts of the country, and discovered great numbers of new plants. He was essentially a descriptive naturalist, and possessed nice power of discrimina-

tion. He published an "Introduction to Botany" (Boston, 1827), which contains a few chapters on the anatomy of plants but is chiefly devoted to the description of the flower and of the Linnæan classification based on its parts. He seems to have regarded botany for the ordinary student as merely a pretty amusement, a notion which his presentation of the subject certainly justifies, and which has not yet disappeared from the public mind. Of his stay in Cambridge the late Dr. A. P. Peabody has written:—

" His name was mythical to the members of the college. We used to hear of him as the greatest of naturalists; but I never knew of his being seen. . . . I think that the catalogue promised instruction by him to those who wanted it; but I never heard of his having a pupil."

Curiously enough, the chief native New England botanist among the contemporaries of Peck and Nuttall was never a teacher of natural history. Many a Bostonian not yet old remembers the charming personality of Dr. Jacob Bigelow. Born in Sudbury, Mass., he was graduated at Harvard in 1806, and from its Medical School in 1810. For forty years after 1815 he was Professor of Materia Medica in the same school, a position for which, says Dr. Peabody, "he had very much the same qualification that a learned unbeliever might have for a professorship of Christian theology. No other man of his time had so little faith in drugs." How much New England patients of two and three generations ago owe to such an influence in the training of their practitioners is beyond calculation. He held also the Rumford chair of Applied Science at Cambridge from 1816 to 1827, and his lectures on this foundation were among the great attractions of the college course. As the chief member of the first committee on the American Pharmacopœia, as the leading spirit in the establishment of Mount Auburn cemetery, and in all good works, he was widely known. A finished classical scholar, he was among the first to depreciate the value of the classics in a practical education, as compared with technical and scientific training. His botanical fame rests on two works. The "Florula Bostoniensis," the first published key to the plants of the vicinity of Boston, appeared in 1814, and new editions were required in 1824 and 1840. It is not easy to overestimate the influence of this classic work in creating an interest in the botany of the widening region covered by its successive editions, or fully to appreciate now the labor involved, and the enthusiasm which sustained it, in the collection of the data for its preparation. The "American Medical Botany," published in three volumes between 1817 and 1820, was long the standard authority on our officinal plants.

Among the contributors of local material for the "Florula" may be mentioned Dr. Andrew Nichols, of South Danvers (now Peabody), a pupil of Dr. Waterhouse, and Dr. George Osgood, of Danvers,—enthusiastic amateurs when such were few.

Contemporary with Nuttall and Bigelow were two botanists of western Massachusetts, who belong equally to New York. Amos Eaton, our first professional teacher of natural history, and especially of botany, was a native of New York state. After experience as a blacksmith's apprentice, he was finally graduated from Williams Col-

JACOB BIGELOW.

lege in 1799. He then studied law, but, after a series of reverses and misfortunes, determined to devote himself to natural history. Pursuing its study, partly at Yale, he began lecturing on botany and geology at Williams in 1817. By this means, combined with the teaching of private pupils and other lecturing, he supported himself until made, in 1824, Senior Professor in the Rensselaer Polytechnic Institute, at Troy, which position he retained up to his death, in 1842. He was the author of various textbooks and of the most widely used "Manual of Botany" of his time. This appeared in its first edition almost simultaneously with the publication of Dr. Bigelow's "Florula"; and its eighth and last edition, called "North American Botany," with Dr. Bigelow's last, marks the year 1840 as the date of the final appearance of the Linnæan hand-books. Professor Eaton was to the end a strong opponent of the natural system which was gradually replacing the artificial one to which he clung, and for which the latter was intended by its author only as a temporary substitute. He was one of the numerous disciples of Linnæus who outdid their master in devotion to his system, because without his insight into the true principles of classification. His conception of the botanist is shown in the following explicit statement of notions diametrically opposed to those of Dr. Waterhouse:—

" No one should ever be employed as a teacher of Botany, unless he can give his pupils at sight the names of at least four hundred species of indigenous plants, growing in the vicinity of his school ; and he ought to be able to recognize from the mere habits of plants six or eight hundred species."

But while this most pernicious feature of Professor Eaton's teaching has been so extensively perpetuated in that of the present, its best aspect has been

AMOS EATON.

as largely lost sight of. Near the end of his life he wrote to teachers of botany:—

" If you have any respect for yourselves, or for human science, I beg that you will never lend your aid in that public imposition which has, within the last dozen years, degraded and debased the study of botany. I mean that of pretending to teach practical botany by school lessons, without having each student hold in his hand a system of plants and living specimens for perpetual demonstration. . . . It is true that pictures may be studied; so may the picture of a blacksmith shoeing a horse be *studied*. But can you become a blacksmith by studying this picture?"

It is only in the last few years that this warning has begun to be less needed than half a century ago.

Chester Dewey was born in Sheffield, Mass., and was graduated at Williams in 1806. After the study of theology and ordination as a minister, a preliminary then thought almost essential to fitness for any college chair, he became Professor of Mathematics and Natural Philosophy in his *alma mater* in 1810, lecturing also in medical schools of the vicinity. In 1850, he became Professor of Chemistry and Natural Philosophy in the University of Rochester, N. Y., and there he died in 1867. His botanical work was done outside of his professional duties, and he is known as a critical student of our grasses and sedges. He also did a part of the botanical work of the Natural History Survey of Massachusetts, preparing the "Report on the herbaceous flowering plants" of the state.

New England claims a large interest in the brothers Boott, "gentlemen of the old school" in the best sense. Born in Boston of English parents, and Harvard graduates, their lives were spent apart. Dr. Francis Boott settled in London, and was for thirty-five years before his death, in 1863, one of its best known practitioners of medicine. He devoted himself with zeal to the elucidation of the great sedge-genus *Carex*, and the classic illustrations of its species published at his own expense are his best and most enduring monument. After he died, his brother, William Boott, a lifelong resident of Boston, "by a sort of *noblesse oblige*," took up the study of the sedges and, though publishing little, became recognized as an authority second only to the elder brother.

The beautiful Connecticut valley inspired and was made famous by the work of one of those "self-made" men who, fortunate in their inheritance and in their environment, have the genius to guide their development to the highest results. After a youth of hard work without opportunity for regular college training, and after a service of

a few years as a country minister,
Edward Hitchcock became, in 1821,
Professor of Chemistry and Natural
History in Amherst College. Here
he served with singular disinterested-
ness to the end of his life, in 1864.
From 1845 to 1854 he was president of
the college, returning to the ranks of
the faculty when the emergency which
had demanded this service was past.
Preëminent as a geologist, he is perhaps best known for his work on the
fossil footprints of the Connecticut valley sandstones. He published in 1829
a "Catalogue of Plants growing
within twenty miles of Amherst College," which may be regarded as the
forerunner of the numerous similar
lists since prepared for various regions.
He was one of the first American
writers to give more than perfunctory
attention to the simpler flowerless
plants. His list of plants of the Linnæan class *Cryptogamia*, although it
now seems meagre enough, shows his
knowledge of the whole field of botany
to have been better than that of most
botanists of the time. He also prepared the first "Report on the Animals
and Plants of Massachusetts," in addition to one on the Geology of the state.

When, after these first reports, the
work of the Natural History Survey
was divided among a number of
specialists, Mr. George B. Emerson
shared the botanical work with Professor Dewey of Williams College,
taking the woody plants as his subject.
Mr. Emerson was a native of Maine,
a graduate of Harvard, and for over
thirty years the principal of the best
private school for girls in Boston. As
a leading spirit in every attempt to give
dignity to the teacher's profession and
to elevate the standard of fitness for its
members, the whole country owes him
a lasting debt. His interest in plants
dates from his acquaintance with Dr.
Cutler's paper in his early youth. To
the preparation of his "Report on the
Trees and Shrubs growing naturally
in the forests of Massachusetts" he devoted all his leisure for nine years.
He personally explored the state from

EDWARD HITCHCOCK.

end to end, despising no source of
information; and the result was a
volume which must remain classic.
It was published at Boston in 1846,
and a beautiful second edition, with
colored plates, followed in 1875. Mr.
Emerson had a genuine love of the
woods, and clearly saw the practical
value to the state of their wise management. At a time when, to the ordinary observer, the supply of timber
seemed limitless, he wrote:—

"Now those old woods are everywhere
falling. The axe has made, and is making,
wanton and terrible havoc. The cunning
foresight of the Yankee seems to desert him
when he takes the axe in hand. The new
settler clears in a year more acres than he
can cultivate in ten, and destroys at a single
burning many a winter's fuel, which would
better be kept in reserve for his grand-
children."

If he felt constrained to speak thus
fifty years ago, what would he say
to-day, when the same reckless policy
which he lamented has brought close
home to us the problem of future
wood supply, and when our stateliest
and most important forests are fast falling into the hands of men as soulless

as their steam saw-mills! It is to Mr. Emerson's book that we owe the beautiful Bouvé collection of woods in the museum of the Boston Society of Natural History; and it was largely through his personal influence that his father-in-law endowed the Arnold Arboretum at Jamaica Plain, now one of the foremost forestry establishments of the world, which is none the less his memorial that it does not bear his name.

One of the most enthusiastic and promising of the New England botanists of two generations ago was William Oakes, of Ipswich. A graduate of Harvard in 1820, he became disgusted with the law after admission to the bar, and thenceforward devoted himself to the study of New England plants. He visited every part of our territory and repeatedly explored the more interesting mountain regions, having planned an elaborate Flora, for which he accumulated an enormous mass of material. But his anxiety for perfection in his work delayed its completion; and finally his death by drowning in 1848 lost to science his

GEORGE B. EMERSON.

accurate and discriminating knowledge. It is related of Oakes that a woman who had watched him for half an hour on his knees near her house, searching for a rare moss, thought him crazy and, in the kindness of her heart, brought him a slice of bread and butter.

His friend and companion on many trips, Dr. Charles Pickering, was of Salem ancestry, a grandson of Col. Timothy Pickering. He was graduated at the Harvard Medical School in 1826, and practised for several years in Philadelphia. In 1838 he sailed with the Wilkes Exploring Expedition to the South Pacific as one of its naturalists, giving especial attention•to the geographical distribution of plants and animals. On the return of the expedition, he made a journey of several years in eastern countries for study. His chief works are one on geographical distribution, published in part by the government, and a "Chronological History of Plants, Man's Record of his own existence," a wonder of learning and of patience, covering 1222 closely printed quarto pages and tracing the migrations and transportations of plants as shown by existing historic records, from the beginning of the first great year of the Egyptian reckoning, with citations in the original lan-

WILLIAM OAKES.

guages. This work was in press at the time of his death in 1878. It has been said, and may well be believed, of Dr. Pickering, that "love of knowledge was the one passion of his life."

In 1833 Harvard College received by the will of Dr. Joshua Fisher, for many years Beverly's beloved physician, an endowment to establish a Fisher professorship of Natural History. After having been declined by Francis Boott and by George B. Emerson, this chair was offered in 1842 to a very promising young man, the pupil and associate of Dr. John Torrey of New York, then the foremost American systematist. This young man who was thirty-two years old and already under appointment to a chair in the nascent University of Michigan, accepted the call, and for forty-six years adorned his chair and the institution of his adoption. The story of Asa Gray's life needs not to be rehearsed here. It has lately been made familiar by the delightful volumes of his letters prepared by his wife. The history of his work is that of American systematic botany for nearly five decades. His text-books and manuals, in various editions, his masterly use of the unrivalled facilities afforded by his position and by the period of exploration and discovery in the great West over which his activity extended, and his delightful personality, combined to give him a preëminence never before accorded to an American botanist and, one may almost as safely say, never again to be enjoyed by any; while the universal verdict of his foreign colleagues places him in the front rank of the botanists of the world. Though he first saw the light in central New York, he was of Massachusetts stock, and, as all his best work was done there, the state may fairly claim him. His first book, published when he was but twenty-six years old, was the "Elements of Botany," and his last, brought out shortly before his death, bears the same title. In 1842 appeared the "Botanical Text-book," and in 1848 the first edition of the

"Manual of Botany." In the latter the Linnæan key was used as an aid to rapid determination; but even this use of it was abandoned in later editions, and the plants were, from the first, arranged according to the natural system of De Candolle. This work determined the disappearance of the Linnæan system from American literature. That Dr. Gray's *chef d'oeuvre*, the "Synoptic Flora of North America," in spite of many years of untiring

industry, was left very incomplete, can never cease to be a source of regret to all who profess any interest in the "science beautiful." It is true that Dr. Gray limited himself, in his own work, to the external structure and classification of the flowering plants; but this field was a much larger part of botany in his youth than at present, and it will yet be long before it fails to afford scope for the exercise of the best abilities.

Two names are closely associated with the work of Dr. Gray at Harvard, those of two natives of Connecticut and graduates of Yale. Charles Wright spent some years after his graduation in 1835 in the exploration of then almost unknown Texas; and the fruit of some of his work was one of Dr. Gray's best known papers, the "Plantæ Wrightianæ." He accompanied the North Pacific Exploring Expedition as botanist, in 1853–5, and spent most of the next twelve years in extensive and fruitful collecting in Cuba. Several years were then passed in research and assistance at the Gray herbarium; and his last years till his death in 1885 were devoted to the exploration of his native hills.

Sereno Watson's love of botany did not assert itself for twenty years after his graduation in 1847. A varied and unsatisfactory experience during this time led him finally to accompany King's exploring expedition in the western mountains from 1867 to 1871, devoting himself chiefly to the flowering plants of the country. The resulting volume on the "Botany of California" showed his grasp of his subject so clearly, and Dr. Gray had been so impressed by his ability during his work in Cambridge while preparing it, that he was offered the curatorship of the Gray herbarium in 1874. The earnest hope of botanists that he might be able, after the death of his chief, to complete the Synoptic Flora, on which they had worked for fourteen years together, was shattered by his untimely death only four years after.

It would be unfair not to acknowledge the debt of our science to the beautiful plates which accompany some of Dr. Gray's publications, as well as often more popular works,

J. L. RUSSELL.

from the pencil of Isaac Sprague. Combining the skill of the artist with the insight and accuracy of the naturalist, his work has set a high standard for his successors.

Two names which here deserve mention as those of contributors of material for Gray's Manual and critical students of the pond-weeds and other flowering plants of our fresh waters are Dr. J. W. Robbins, Yale, '21, a physician of Uxbridge, Mass., and Rev. Thomas Morong, Amherst, '48, who was settled over several Massachusetts parishes and late in life carried out a two years botanical exploration in South America. His last years were spent as curator of the herbarium of Columbia College, in New York.

Under the Linnæan system the flowerless plants were thrown together into a class coördinate with each of the classes of flowering plants, and placed at the end of the list. This treatment was perhaps natural enough in view of the scanty knowledge of these forms then possessed; but it had the effect of making them regarded as of little consequence, and for a long time almost ignored. Up to 1840 their study in America had been fragmentary in the extreme. But with the advent of natural classifications which first indicated something of the true place of the different cryptogamic groups in the vegetable kingdom as a whole, they began to attract the attention of lovers of plants. Perhaps our earliest student of the simpler aquatic plants included under the general name *Algae* was Jacob Whitman Bailey, a native of Worcester County, Mass. A graduate of West Point, he soon became Professor of Chemistry, Mineralogy and Geology there, and up to his early death in 1857, contributed much to the knowledge of his favorite plants. He remembered his New England origin by leaving his fine collection of microscopic preparations to the Boston Society of Natural History.

The first systematic attempt at an account of American algæ was left to an Irishman, Prof. William Henry Harvey, of Trinity College, Dublin. An enthusiastic collector and traveller, as well as a noted authority on these plants, he had already resided several years in South Africa, when, in 1849, he visited the United States to investigate its marine flora, spending considerable time in New England. The result was a work issued in three parts during the fifties by the Smithsonian Institution, —the "Nereis Boreali-Americana," which is still indispensable in the study of our "sea-weeds."

A pioneer of much greater ability than his published work shows was a Unitarian clergyman of Massachusetts, Rev. John Lewis Russell. A devoted lover of nature, his study was always a naturalist's den, full of "green things and beautiful." He was fond of giving popular instruction, whether by voice or by pen, and was a prominent botanical contributor to the "New American Cyclopedia." For forty years, and till the close of his life in 1873, he was "Professor of Botany and Vegetable Physiology" to the Massachusetts Horticultural Society; and he left to it a bequest to endow an annual lecture on the relations of the fungi to horticulture, as though he foresaw the important discoveries since made and yet to be made concerning the etiology and preventive treatment of the numerous and destructive fungous diseases of our cultivated plants. His special studies upon the mosses and lichens were begun when very little was known of these plants in America, and it is to be regretted that his extensive knowledge of them was not preserved

to us. It fairly characterizes him to say that "if his personal ambition had been greater, he would have attracted more notice from the world."

But a few years younger than Mr. Russell was the man to whom we owe our chief knowledge of the American lichens, Edward Tuckerman. A member of a well known Boston family, and a Harvard man by tradition, he was sent to Union College, where he was graduated in 1837. Ten years later he atoned for this break in the

SERENO WATSON.

family record by taking the A. B., in course at Cambridge, having been earlier graduated from the Law School there, and later completing the course in the Divinity School. Meanwhile he had studied in Europe and had learned from the father of systematic lichenology, Elias Fries, of Upsala, the traditions and methods of that study. As a young man he was an enthusiastic explorer, and the grand "Tuckerman's Ravine," on Mount Washington, perpetuates the memory of his energy. He was a profound student of history, and considered that his professional work, with botany as his recreation. Yet it is as a botanist that he is and will be best remembered. From 1854 to 1873, he was Lecturer in History in Amherst College, in whose shadow his home was made until his work was finished in 1886. From 1858 until the end he was also Professor of Botany there. The volumes he prepared will long remain the chief authority on our lichen-flora. He clung, to the last, to the belief of the older lichenologists in the autonomy of these plant-combinations, a view now superseded; but this detracts nothing from the value and critical

insight of his systematic studies. His "Catalogue of Plants growing within thirty miles of Amherst College" was an elaboration and revision of President Hitchcock's earlier one. For him, as for his predecessor, *plants* meant *members of the vegetable kingdom*, and his list is as complete as possible in all groups. The list of *Fungi* was prepared by the shoemaker botanist of Brattleboro, Vt., Charles C. Frost, one of our earliest students of this group of plants, in whose study he rendered real service. Entirely self-taught, he gained command of several languages

to fit himself for his studies, and was an interesting example of what persistence inspired by devotion can accomplish.

One of the earliest botanical amateurs in Rhode Island was Stephen T. Olney. Though engaged in business during most of his life, he found time for much botanizing. Indeed his fondness for this form of recreation gave currency to stories of mental derangement, which were seriously listened to by the learned court concerned in the settlement of his estate. So well is the love of nature appreciated by the gazing crowd! After studies of the flowering plants in his earlier years, which resulted in the publication of his "Catalogue of Rhode Island Plants," he became a careful investigator of the sea-weeds, a list of which, published a few years before his death in 1878, was entitled "Algæ Rhodiaceæ."

One whose residence among us during the last thirteen years of his life demands a place for him here is Thomas P. James. A wholesale druggist in Philadelphia until 1869, he then removed to Cambridge and devoted himself to the study of his favorite plants, the mosses. The chief result of his labors was the "Manual of the Mosses of North America," left incomplete at his death, and finished, as a labor of love, by his friend, Sereno Watson.

The æsthetic and scientific interest of the ferns has gained for them many appreciative students, one of the chief of whom was a son of New England. Daniel Cady Eaton was graduated at Yale in 1857, and spent his life, just closed, in her service as Professor of Botany, his love for which was, doubtless, a natural inheritance from his grandfather, Amos Eaton. His elaboration of the account of the ferns for successive editions of Gray's Manual, and various other publications, both technical and popular, concerning them have given him the first place among our authorities on this group. He has also done good service in the study of marine algæ and of the general flora of his vicinity. The "Catalogue of flowering Plants and higher Cryptogams growing within thirty miles of Yale College," issued by the Berzelius Society, shows everywhere the influence of his advice.

Certain societies deserve mention here for their influence on the development of botany in New England. The New England Society for Promoting Natural History, formed in December, 1814, became, a month later, the Linnæan Society of New England. From 1818 till its dissolution in 1822, its meetings were very irregularly held; and then, for several years, the outlook was felt to be too discouraging to warrant another attempt. But, in 1830, the Boston Society of Natural History was formed, and, in the next year, incorporated. At that time there was no good museum and no good teaching of natural history in New England. Through aid from the state and private gifts, the Society's fine building on the Back Bay was obtained, and gradually its splendid collections have been accumulated. Its influence on teaching has been steady and important, and its publication fund has furnished the means for giving to the world many scientific memoirs of the greatest value. The Natural History Survey of Massachusetts, authorized through its influence and made under its supervision, was the first in the country and the model for subsequent similar undertakings. Its first active president, who served for seven years, was Dr. Benjamin D. Greene, of Tewksbury, a devoted amateur botanist, who, though publishing nothing, gave freely of his ample means to the society and otherwise for the promotion of science, and left to it his valuable library and herbarium. His successor for six years was George B. Emerson.

The American Academy of Arts and Sciences has counted among its members our ablest botanists, and its publications contain much of the fruit of their labors.

The Essex County Natural History

DANIEL C. EATON.

Society had for its first president Dr. Andrew Nichols. For three years before 1848, Rev. J. L. Russell stood at its head; and after its fusion with the Essex Historical Society, in that year, to form the Essex Institute, he served the last as vice-president for thirteen years. Several of his too brief papers are among its publications.

As one glances over the list of those New Englanders who have added something to our knowledge of plants, he must be struck by the small number of professional botanists it comprises. It is safe to say that only three or four of all those above mentioned have derived their living principally from their botanical work. But we have seen that physicians, clergymen, general naturalists, chemists, teachers, men of business and of leisure have earned for themselves the name of botanist. Our first professional naturalist to devote himself exclusively to botany was Dr. Gray; and it has been only with the development of the newer theoretical and practical aspects of the science that such a career has become more generally possible. In the limits of a single article it will be impossible to give any detailed account of these modern points of view or of the work of living botanists. The activity of recent years is therefore very inadequately represented here. It will be seen that the work of those of New England's botanists whose labors are ended has been that necessary in a new country, the work of exploration and description, opportunities for which are still far from exhausted in our own continent.

It is now a little more than thirty-five years since the dawn of the new botany with the publication of the "Origin of Species," in whose great author we of English blood and language claim a peculiar interest. The new conceptions of development in accordance with determinable laws, of adaptations to determinable conditions, and of real blood-relationship between plant species and genera, which we owe to Darwin and to whose ready appreciation by American botanists Dr. Gray contributed so much, gave, indeed, a new zest and a new interest to descriptive studies. But in opening up innumerable problems before unthought of, in leading to points of view before unreached, it made of the dead science of animals and the dry science of plants one quickening science of life. Since botany and zoölogy were thus brought together, the time has never been when any but a great genius could be master in both fields. Students of plants and animals have both contributed to the solution of biological problems; but it is increasingly true that one must be an animal biologist or a plant biologist. Most of the higher institutions of learning in New England now recognize the modern biology, though some have very tardily done so. Yet of these none but Harvard gives any adequate representation to the botanical aspects of the subject. The greatest service which can now be rendered to the cause of biological science in our midst is provision for teaching and research in the colleges on botanical lines, as it is already conducted by zoölogists.

A HISTORICAL SKETCH
OF THE DEVELOPMENT
OF BOTANY
IN NEW YORK CITY

Henry H. Rusby

Vol. 6   No. 6

TORREYA

June, 1906

A HISTORICAL SKETCH OF THE DEVELOPMENT OF BOTANY IN NEW YORK CITY.*

By Henry H. Rusby.

It is my purpose this afternoon to direct your attention to the influences whose workings have brought into existence the present highly satisfactory organization of botanical work in this city Among many minor elements, three stand out prominently, and call for our special attention. They are : (1) local botanical gardens, including the present one, and the persons who have been associated in their management; (2) the botanical department of Columbia College ; (3) the Torrey Botanical Club.

Were we to commence with the very earliest botanical history of our city, we should be carried back to a time when, as an important seaport in a new world, it was made the temporary headquarters of visiting botanists, who accumulated here their collections, maintaining some of them in a living condition, until the arrival of a convenient opportunity for dispatching them to the mother countries. Such occurrences as these, exerting little influence in the permanent development of a botanical center here, occupy no place in to-day's consideration. Developmental work of the kind that concerns us was active, previous to the close of the 18th century, at some points farther south, especially at Philadelphia, and in New England, but not at New York.

The first important event here was the work of Doctor, afterward Governor, Cadwallader Colden and his daughter Jane, who, near the middle of the 18th century, conducted their studies with the aid of a small botanical garden at their home, near Newburgh.

* An address delivered before the Torrey Botanical Club at a special meeting held on May 23, 1906, in commemoration of the tenth anniversary of the commencement of work in the development of the New York Botanical Garden.
[No. 5, Vol. 6, of Torreya, comprising pages 81–100, was issued May 23, 1906.]

Perhaps the most important part of this work consisted of the correspondence carried on with native and foreign botanists regarding their local flora, and the transmission of specimens. Miss Colden first made known our pretty little *Coptis*, or gold-thread.

A much more important event was the arrival here, in 1785, of the elder Michaux, who established a celebrated botanical garden at New Durham, N. J., the site of which is now occupied by the Hoboken cemetery. A brief account of this garden may be found in the *Bulletin* of our Club, **11** : 88. 1884. In that year I saw growing there a barberry bush which apparently represented the last trace of Michaux's plantings, except that the European medicinal shrub *Rhamnus Frangula*, which he appears to have introduced, has established itself in the adjacent lowlands, and at some neighboring points. Michaux's garden was established especially for the temporary cultivation of plants designed to be sent to France, or to yield seeds designed for such shipment. Nevertheless, so zealous an investigator as Michaux could not fail to utilize this agency for purposes of study, and his great work, *Flora Boreali-Americana*, published in 1803, and other works on North American botany, were thus materially enriched. Michaux's work in this country was continued by his son, one of whose important publications was a *Histoire des arbres forestiers de l'Amérique Septentrionale*, afterwards translated into English as *The North American Sylva*, and this also profited largely by the observations made by the father, while maintaining his garden.

During the time when the Michauxs were so active here, Mr. Samuel L. Mitchill was assiduously collecting plants in the vicinity of his home at Plandome, Long Island, a catalogue of which was published in 1807. His work is of special interest to us, since he was the first professor of botany in Columbia College.

The flora of Manhattan Island was at this time being very actively studied by Major John Le Conte, who in 1811 published an important catalogue relating thereto.

It is a well recognized historical fact that up to this time, and indeed for a long period following, botanical work proper in this country, consisted chiefly of the collecting and naming of plants, and the description of new species.

Writing of the period about 1814, made memorable by the publication of Pursh's *Flora Americae Septentrionalis* and Bigelow's *Florula Bostoniensis*, Darlington says "Botanical works now began to multiply, in the United States — and the students of 'the amiable science' found helps in their delightful pursuit, which rendered it vastly more easy and satisfactory than it had been to their predecessors."

The next botanical undertaking in this city was of the greatest importance in connection with our study, and calls for our particular attention. The successor of Dr. Mitchill as professor of botany and materia medica in Columbia College was Dr. David Hosack, a man of equal breadth and of great strength and energy. His interest in botany was chiefly medical. Most of the amateur botanists of that day were practising physicians, and many, if not most of the professionals had received a medical education and training, so that Dr. Hosack's attitude toward the science was not at the time peculiar. This fact reminds us that outside of the investigation of general and local floras, in their relations to geographical and taxonomic botany, interest then centered chiefly in the medicinal properties and uses of plants. A comparison between this branch of study as then understood and as now conducted can be briefly placed before you by stating that most of the plants then regarded as the important medicinal agents have been dismissed by modern medicine, except where it is trammelled by medical sectarianism. The explanation of their error is not that their results were reached empirically, for this is an excellent method, but that their empirical processes were full of natural sources of error, depending on impressions produced upon unqualified observers, among both patients and practitioners. The chemistry of plants was then practically unknown, whereas it is now the basis of medical botany. Since chemistry constitutes at the same time the visible basis of physiology, and physiology brings us as close as it is possible for us to get to the life of the plant, it follows that medical botany, while not entitled to the objective position that it held in the days of Hosack, is concerned with the same phenomena which engage the attention of the very highest workers in botanical science at the present day.

The great difference between the latter and the work as pursued by Hosack lies in our knowledge of the nature of the life processes and therefore of the proper and effective methods of studying them. Even in the state of ignorance which then existed, it was clear to such keen reasoners as Hosack that the reaching of sound botanical conclusions required that the living plant be kept under observation, and he became possessed of the strongest determination to establish a botanical garden adequate to the needs of local botanists and teachers of botany. After long efforts to secure sufficient coöperation, he at length decided to enter independently upon the enterprise, and in 1801 he purchased 20 acres of land at Elgin, now bounded by 46th and 50th Sts., and 5th and Madison Avenues (or probably of somewhat greater extent) and established the famous Elgin Botanical Garden, better known perhaps as the Hosack Botanical Garden. Besides his hardy plants, many were grown in a large conservatory. The site of this garden was described in 1811 as "about three and one-half miles from this city, on the middle road between Bloomingdale and Kingsbridge." This garden has of late years become so well known through various writings, that I shall not take up its general history. Hosack announced its primary object of attention as being the collection and cultivation of the native plants of this country, especially such as possessed medicinal value or were otherwise useful. He gratefully acknowledges assistance received in starting his Garden from Professor Mitchill, his predecessor, from the Hon. Robert R. Livingston and from John Stevens, Esq., of Hoboken. He soon learned what has recently become apparent to many persons here present, that the successful conduct of a botanical garden is a work of enormous labor and serious responsibility, and that one man, otherwise engaged, cannot accomplish it. With the garden already in actual successful operation, it was not so difficult to enlist state interest, and the legislature was induced to purchase it in 1810, and to provide the necessary funds by means of a lottery. Hosack subsequently enjoyed the classical distinction of all successful promotors of great enterprises, in being assailed by the high-class scum of citizenship. By subsequent legislative action the property was turned over to Columbia College, and its use diverted from that of a botanical garden to that of highly profitable rentals.

We cannot understand the botany of Hosack's time without a brief glance at some of his contemporaries and immediate successors, especially those who exerted local influence. The list includes the names of some of the most honored of American botanists. Biographical sketches of·all are to be found in our *Bulletin* file, so that I need not repeat the purely historical data, but may speak of the character of these men and of their work, in its relation to our subject. Foremost of them all was John Torrey, whose name is commemorated, I hope permanently, in that of our society. Following Dr. Hosack, he was the third of the five men who, up to the present, have occupied the chair of botany in Columbia College. His characteristics may be expressed in the terms, strong personal character, broad scholarship and great intellectual ability. Although best known to us as a botanist, yet thirty years of his life were those of a great teacher and worker in chemistry at the U. S. Military Academy at West Point, in the College of Physicians and Surgeons of this city, in Princeton College, and as U. S. Assayer in the New York office. Had the necessary facilities then existed in this country, it seems likely that this man, combining such a great knowledge of botany and chemistry, might here have developed important researches in the chemistry of plants. As a matter of fact, his knowledge of botany was acquired chiefly as a recreation in the hours of leisure afforded by his other professional work. Yet Underwood truly writes, "When the annals of American botany are finally written, no name will have a more conspicuous position than that of John Torrey."

Almost before reaching manhood Torrey was one of the founders of the New York Lyceum of Natural History, and was the leader in publishing, through it, a catalogue of plants growing within thirty miles of the city. Five years later he published the first part of his *Flora of the Northern and Middle Sections of the United States*, and later his *Compendium* on the same subject, important forerunners, in more than one way, of Gray's Manual.

These accomplishments proved him the great master that he was, and soon his hands were crowded with important work, especially connected with the active explorations of our western territory then in progress. In this work he was a close associate of Asa Gray, and probably their most important work was the first parts of their *Flora of North America*, published from 1838 to 1843. Many men whose work has thus branched out from local into general lines have allowed the latter to supplant and replace the former, but this was not true of Torrey, who combined in rare degree generic and specific powers. Not only were his interest and activity in local work undiminished, but they grew apace, and his patient and quiet enthusiasm gathered about him a group of associates who not only were devoted to him personally, but imitated and emulated his work. In this saying is stated the immediate origin of the Torrey Botanical Club. At various points in the history of our Club, we have been reminded that "a nation has arisen that knew not Joseph," and various proposals have been made for changing the name of the society. Let us record now the opinion that the selection of Torrey's name for this purpose was so just, natural and appropriate that its retention amounts to a historical necessity.

Except for the published works of Torrey, most of those of this early period which here concern us were of a somewhat general nature, but naturally including our local interests. Of these may be mentioned the following: In 1813, Muhlenberg's *Catalogue of North American Plants*, and in 1817 his work on North American grasses and sedges; in 1818, Nuttall's most scholarly work on the genera of North American plants; in 1820, Gray's *Genera;* in 1822, Schweinitz's *Monograph of the Genus Viola;* in 1833, Beck's *Botany of the Northern and Middle States;* in 1834, Schweinitz's work on North American Fungi, and in the same year, Gray's *Monograph of the North American Species of Rhynchospora*. In the meantime, very important works of a similar character were being produced in the South, and to a lesser extent, in the West.

These publications, it will be observed, were chiefly of interest to those actively engaged in original work, and not to young students. In 1803 there appeared about the first work designed especially for the latter class, an elementary work on botany by Barton. Writing of 1824, Darlington says: "About this time some of the schools in the Northern States began to make a profession of teaching botany, and a demand for suitable books for this purpose arose. Accordingly, a number, such as they were, soon appeared. Among the most successful was a *Manual*, compiled by Professor Amos Eaton, of Troy, New York." Of the character of the educational works of the period, little need be said, since it is sufficiently indicated in that of the work in which botanists were then engaged. This sort of botanical teaching entered upon its most active stage with the appearance of Gray's *Elements of Botany*, in 1836, a work that is still being sold upon an extensive scale, and this, in your speaker's opinion, very greatly to the advantage of botany, in spite of the many books of different character, the use of which we so greatly enjoy. The publication, for the use of students, of text-books on structural botany, and later on morphology, in connection with manuals on local floras, became very popular, and of incalculable value in interesting people in the study of plants.

We must now pass from this general consideration of local botanical development up to the middle of the last century, and follow some special influences proceeding from the growth of the botanical department of Columbia College. During the period when Dr. Torrey was at its head, that department was very actively engaged in educational work, though this was of the peculiarly restricted sort characteristic of the times. About the time of his death in 1873, his herbarium and library, which he had previously maintained in his home, came into the possession of Columbia, together with the herbaria of Crooke, Chapman and Meissner. To these, collections from various parts of the world, and especially from those parts of the United States then being explored, were rapidly added, and a very large and important herbarium soon grew up; but no professor of botany was appointed to succeed Dr. Torrey, and the herbarium was neglected by the curator in charge. A very large part of it was not classified, nor even named, and lay in the form of a small mountain of dusty bundles which were not, and could not be consulted. Botanical instruction was most meager, and was merely a part of the general course in biology. There was not, in fact, a department of botany, the subject being treated as a subordinate of geology, under Professor John S. Newberry. From 1875 to 1879, Dr. Britton was a student at the School of Mines, and was strongly attracted, by natural taste and ability, toward the botanical side of his work. When upon his graduation he was appointed assistant to Dr. Newberry, he appreciated clearly the great value of the materials for a botanical department, to be organized on a new and modern basis, which were in the possession of the College, and he began a careful and systematic examination of them. In speaking of this exceedingly important event in the general, as well as in the botanical, history of New York, your speaker takes the keenest delight, as he was for most of the time one of the closest associates of Dr. Britton, and can speak of that which he not only saw, but which he watched with appreciative interest.

A special stimulus to Dr. Britton at this time was his interest in his first great botanical undertaking, the preparation of an elaborate catalogue of the plants of New Jersey, this also, being performed subordinately to a department of geology. In this undertaking, an intimate association with the members of our Club and an active participation in its work were prime essentials to success, an illustration of the way in which existing forces worked together in carrying forward our natural botanical development. Another potent influence of a similar nature should be here recorded. At this time considerable botanical material from distant parts of this country and from other hitherto unexplored regions was coming to this city for original study, and this made it imperative that Columbia's botanical house should be set in order in the interest of comparative work. With the knowledge and encouragement of Dr. Newberry, but with comparatively little on the part of others concerned in the management of the college, Dr. Britton carried on this work in the interim of his official duties, until at length a great working herbarium existed where before there was chaos. At the same time the botanical instruction was being extended and, of greater importance, was being modernized. When the Doctor was at length prepared to make the situation known to Columbia, it was not to submit plans for the organization of a botanical department, but to present to it one already made, and requiring only to be officially recognized and formally named. The performance of these ceremonies, with suitable provision for maintenance, guaranteed the position of New York as one of the first botanical centers of the country, and later of the world, with Dr. Britton as Columbia's fourth professor in this department. Thus we see that at every important stage in its development, the botanical department of Columbia has owed its prosperity not to the institution as such, but to some earnest worker, ready to make the sacrifice of love. Hosack individually made the botanical garden that afterward enriched the institution; Torrey accumulated the herbarium that became the corner-stone of the later structure ; Britton silently — one may almost say surreptitiously — brought about changes which have finally placed it in the vanguard of the world's botanical forces.

The intercourse and personal and professional associations dependent upon the increasing number of persons in and about New York who became interested in botanical work in Torrey's time led most naturally and inevitably to a botanical society, at first incidental and unorganized, later a formal organization.

As is true of so many institutions which grow healthily and attain to great and permanent success, the exact date of the origin of our Club can hardly be fixed. Those of you who take even the slightest general interest in this subject should not fail to read * the inaugural address of Dr. George Thurber, delivered at the Astor House in 1873, on the occasion of his first election as our first president. He confesses his entire inability to fix on the time when Torrey and his friends virtually established the society. He says that for a long time after the election of the first set of officers the members found it impossible to break from the habit of informal, free-and-easy, conversational meetings which had grown up and which, I must remark, have always been found the most effective in the Club's work, whenever they have recurred.

* Bull. Torrey Club, **4** : 26-39. 1873.

The Club's formal organization was undertaken in 1867, and its incorporation occurred four years later, under the name New York Botanical Club, changed the following year to that which it now bears. Within three years after its establishment the Club began issuing a monthly publication, the *Bulletin*, since uninterruptedly maintained. Its prefatory note declared its primary object to be "to form a medium of communication for all those interested in the Flora of this vicinity, and thus to bring together and fan into a flame the sparks of botanical enthusiasm, at present too much isolated. . . . We have chiefly in view the development of a greater botanical interest in our neighborhood, and found our hopes of success as much upon learners as upon the learned." May I pause here to ask all those present to regard this sentiment as that which actuates our Club to-day. There have been unfortunate periods in our history when this fundamental principle has been lost sight of; when learned newcomers, unfamiliar with our history and character, have assumed that we existed for the learned only. Believe me that this spirit does not exist to-day. We are most desirous that the knowledge should go abroad that the Torrey Botanical Club exists and is maintained for the most humble learners, equally with the learned, and our invitation to membership is to-day most cordially extended to everyone who desires either to assist in strengthening our influence, or to be assisted by us.

In the further unfolding of its objects, the *Bulletin* unconsciously states the object of the Club's organization : "An attentive study of plants in their native haunts is essential to the advance of the science, and in this respect the local observer has an advantage over the explorer of extensive regions, or the possessor of a general herbarium. He can note the plant from its cradle to its grave; can watch its struggles for existence, its habits, its migrations, its variations; can study its atmospheric and entomological economies; can speculate on its relations to the past, or experiment on its utility to man." Ecology is thus clearly seen to be the object of study, notwithstanding that the name of it was not generally discovered by our botanical fraternity until about 1890, nor the active and merciless chase of the poor thing by American botanists well under way until about five years later.

From this time up to the establishment of the New York Botanical Garden the history of our Club is practically that of botany in this city, for very little was done that was not directly or indirectly connected with us or, one might say, actually centered about us. This fact is of the utmost importance in our study, since upon it depends the essential character of most of what has since occurred.

The Club's history is so voluminous that it requires separate and extended treatment, and I can here do little but refer to its influence. Its first officers were George Thurber, president; Timothy F. Allen, vice-president; A. A. Crooke, treasurer; James Hogg, corresponding secretary; P. V. LeRoy, recording secretary; William H. Leggett, editor; P. V. LeRoy, curator.

Some of the more influential of the early members call for attention at this point.

(To be continued in the July number)

Vol. 6   No. 7

TORREYA

July, 1906

A HISTORICAL SKETCH OF THE DEVELOPMENT OF BOTANY IN NEW YORK CITY

By Henry H. Rusby

(Continued from page 111)

Dr. Thurber, our first president, was characterized by profound conscientiousness and great determination. He began life as a pharmacist, in Providence, and developed a strong leaning toward chemistry, of which subject he became a teacher. His love of botany grew out of his study of drugs. In 1850 he went as botanist, quartermaster and commissary to the Mexican Boundary Commission, the botanical results of which were published by Torrey in 1859. He received the degree of A.M. from Brown University, and the honorary degree of M.D. from the University Medical College, of this city. He was in the U. S. Assay Office for two years and left from motives of honor. He was at various times a teacher in Cooper Union, the New York College of Pharmacy and Michigan Agricultural College, and was president of several horticultural societies and of this Club until 1880. For twenty-two years he was editor of the *American Agriculturist*, in which capacity he exerted an influence over the character of young people, in the agricultural sections of the country, that was and is of great national importance. His most important contribution to botanical work was perhaps the maintenance of a botanical garden at Passaic, New Jersey, in close relations with that of Harvard. His private fortunes were melancholy. Captured by the whirl of speculation in real estate that followed the civil war, he purchased land at an excessive price, and spent the rest of his life in a painful struggle honorably to discharge his financial obligations.

[No. 6, Vol. 6, of TORREYA, comprising pages 101–132, was issued June 20, 1906.]

133

Mr. Wm. H. Leggett, our editor until near the time of his death in 1882, was a distinguished and successful educator, maintaining a private school in the upper part of the city. He was described as a "profound classical scholar," making a specialty of Greek. Notwithstanding this predilection, he managed to perform his botanical work in a most creditable manner, and exerted a persuasive influence in interesting the young in this study. It must not be overlooked that in founding our *Bulletin* he assumed the financial responsibility for its success.

Professor Alphonso Wood will be ever remembered by American botanists as the author of descriptive floras of the highest scholarly character, and put together with a rare regard for educational principles. Those who are fortunate enough to have owned and carefully used his books will recognize, in the light of our present advancement, that his knowledge of plants was more full and accurate than that of most of our American botanists who have written similar works. His life was not a happy one. The influences of prestige and station were deliberately turned against him, and he was to a great extent suppressed. The manuscript of his Class-book was used by him in teaching, and steadily perfected, for ten years before its publication, which was very successful. His work in life was that of an educator. He taught in and presided over a number of institutions, and brought educational and financial success wherever he went. In 1865 he made an overland botanical journey to California, then to Puget Sound, and home by way of the Isthmus. The specimens and observations accumulated on this journey were very valuable, but have never been systematically studied. He was professor in the New York College of Pharmacy during the two years preceding his death, in 1881.

Mr. Coe F. Austin was born at Closter, N. J., in 1831, and died in 1880. His chief characteristics were a marvelous energy and capacity for work, and great independence and originality in selecting his lines. His energy was closely confined, so far as general botany was concerned, to the local flora, and no other man has done so much to make known the flora of northern New Jersey. He was at the time one of the very few local workers in bryology and practically our only close student of the Hepaticae. Unfortunately, his botanical zeal caused his family to be deprived of many of the important possessions of this life.

Mr. M. Ruger, who died in 1879 at the untimely age of 44, was in many respects a memorable character. His physical constitution was so weak that he could never attend school, nor engage in any vocation, yet he succeeded in acquiring a very liberal education, and in pursuing the avocation of botany until he came to be known as the Club's "walking encyclopedia." His knowledge of the local flora was remarkably full and remarkably accurate, and before he died this knowledge was extended over a large part of the country. Not only did his observations enrich the proceedings of the Club and the pages of the *Bulletin*, but his collections did much to build up the Club's herbarium. His work was notable for extending into such fields as that of mycology, then almost unworked, and for many years all questions arising in the Club relating to fungi were habitually referred to him. He was stricken down while botanizing and died two days later.

Professor Joseph Schrenck was a school principal in Hoboken, who applied his scholarly tastes and abilities to the study of botany in ways then little known among us, and he labored diligently and with great patience to lead others in the same direction. He obtained a professorship to do evening work in the College of Pharmacy. This work, along strictly technical lines, led him to a deeper study of plants, both anatomical and physiological, by the use of the microscope and chemical reagents, than that which then prevailed here. From this experience he was soon led to deplore the superficiality of current work, and he started private classes among the Club's members for interesting them in methods which he saw must soon become dominant. Although general tendencies were not thus changed, many persons were interested, and some of our best workers of the present day acquired their first training in this direction from these humble efforts of Professor Schrenck.

During the same time another worker, Professor E. H. Day,

who reminds us of Schrenck in some ways, was active in similar work at the City Normal College. Tied down by the unceasing drudgery of wholesale elementary teaching, he might have been pardoned for falling into the rut and then into the slough, but on the contrary, he kept both his interest and his activity fresh, and he was ever alert in inspiring his students with a love of the subjects studied, which might lead them later to continue their studies as amateurs. In 1883, while occupying the chair at a Club meeting, he suggested the appointment of a sub-section for the study of physiological botany. A committee was appointed, consisting of Messrs. Hyatt and Britton, and Miss Knight, now Mrs. Britton. This was perhaps a very important historical event.

Dr. Timothy F. Allen had one of the longest uninterrupted careers as a member in the annals of the Club, extending from its foundation to 1902. During the early part of this career he was very active in the meetings and in all the work of the Club, and later he developed an interest as a successful investigator of the Characeae. His later life was an intensely busy one in the field of medicine, both as a practitioner and teacher, and his botanical activity was to a great extent crowded out, but he never lost his interest in the Club, nor did he ever fail in his readiness to respond to any special call for coöperation.

Mr. Wm. H. Rudkin was an active down-town business man, who lent his fine abilities to the financial management of the Club as its treasurer for many years when this duty required faithfulness, tact, sacrifice and responsibility. He was by no means wanting in botanical acumen, nor failing in activity, but it is in the capacity above mentioned that he is to-day deserving of our special remembrance and gratitude.

Dr. Emily L. Gregory, though not one of the older members of the Club, exerted a profound influence upon its character and upon that of botanical work in the city. Thoroughly educated in the best modern schools of Germany, and especially a disciple of Schwendener, she became here a missionary of advanced work and methods. She founded the botanical department of Barnard College and established there a botanical center which has since steadily grown in strength and influence, and is now one of our most important botanical possessions.

It has been seen that the work of the Club was at first narrow as to the subjects involved, because the science itself was so, especially in this country. It continued afterward to retain this character, largely by force of habit. It is not true, however, as has been generally accepted, in response to the criticisms of those who did not know, that its work was confined to accumulating and naming specimens, enumerating circumscribed floras and studying individual structures. Its work was the study of living manifestations of plants in the field, a study which has of late been largely eliminated, to the very great misfortune of science, as here pursued. There came a time when New York experienced an invasion of botanists with concepts, knowledge, interests and methods which were largely foreign to us. Their importations were of incalculable value to New York, and at the same time most urgently needed, and resulted in giving to us a new, modern and broad botany. The event was not, however, free from unfortunate incidents. Laboratory work was given undue prominence. Ecological and other field work came to be largely neglected, and what might not inappropriately be called the disjointed period of the Club's history ensued.

With a few closing remarks, the history of the Club must be dismissed from further consideration. Its publication work has steadily increased, until it now includes three periodicals, the smallest much larger than was the *Bulletin* until many years after its commencement. It has published catalogues of plants of local and distant areas, monographs of important groups, and results of important anatomical, physiological and economic researches. It has collected lists of works and workers, and maintained indoor scientific meetings, at first one, then two monthly, and delightful, and on the whole, very profitable, field meetings, hereafter to be conducted on a systematic basis not previously attempted. It has conducted elementary courses of instruction, and given lecture courses. Its work has included every part of the vegetable kingdom, and covered almost every part of the world. Its influence in securing the establishment of our present botanical garden may next be considered.

So eager was the desire of the early members of the Club to observe how plants lived, that many of those able to own gardens ignored vegetables and flowers, and maintained little botanical gardens at their homes. Mr. Wm. Bower, for example, was a hard-worked die-cutter of Newark, yet he managed to accumulate, in his little city yard, a choice collection of native and foreign rarities. These statements relate to a period when the most generous botanizing grounds were still within easy reach of everyone, some of them existing even in the heart of the present city.

As succeeding decades of extending settlement destroyed the localities which had been so greatly prized, not only in the remote parts of the island but in the country round about, these people not only mourned their present loss, but were alarmed by the handwriting on the wall, and the demand for a botanical garden arose independently in the mind of every botanist, professional and amateur. So early as 1874 the Club appointed a committee to act with the New York Pharmaceutical Association in requesting the city to establish such a garden in Central Park.

As the educational side of our work grew in importance, and especially in breadth, and as the student body doubled and redoubled, the cry for the garden grew equally loud from that direction, and continued until at length it was satisfied. The great value to Harvard and its work of the well-managed plot that it utilized in this way was appreciated and often discussed at the little meetings which gathered around the old pot-stove in Professor Newberry's room, during his presidency of the Club.

Under the influence of Columbia's progress, as already described, it appreciated this want as much, probably, as any other of our botanical elements. Its peculiar relations to the former Elgin Garden were recalled in the public press. A contributor to the New York *Herald*, of November 26 and 27, 1888, made an earnest appeal for the recognition by the city of this great want. Dr. Arthur Hollick, to whose faithful and self-sacrificing work as secretary, our Club largely owed its strength for a prolonged period, directed our attention to these articles and proposed that he write an official letter to the *Herald* endorsing them. Such a letter was authorized, and it appeared on December 2 following. A committee was appointed consisting of Dr. Hollick, Mr. E. E. Sterns, and Professor Newberry, to deliberate and report to the Club whether it were advisable for us to take any action for the furtherance of this movement. The possibility of the realization of our long cherished hopes now began to take possession of our minds, yet without any very strong hope being entertained. The Club had no political influence and little acquaintance with those financial interests, the aid of which was rightly deemed to be essential to success. As it resulted, however, some of these men were led to interest themselves in the proposition, largely through the influence of Judges Addison Brown and Charles P. Daly, and of Mr. Charles F. Cox and Mr. Wm. E. Dodge. For a long time the idea was regarded with favor in influential circles, but without any definite steps being taken to execute it. Finally, it was remembered that all history teaches that when you have wearied of discussing a project, and are at length really resolved to carry it out, you must call in the assistance of the women. So a ladies' committee was appointed and held a memorable meeting at the residence of Mrs. Charles P. Daly, which some of the men, your favored speaker among them, were graciously permitted to attend. This influence, while but one of many, each of which was necessary to success, seemed to give the final impetus needed. Mr. Cornelius Vanderbilt assumed the financial and executive management of the enterprise, and the stage of organization was reached.

One element in the success of the Garden that has already shown itself to possess a value beyond price, and which is certain to do so with increasing clearness in the future, is the protective influence of its charter. Born of the learning, long and wide experience and ripe judgment of Judges Brown and Daly, and occupying their attention for considerably more than a year before they were willing to regard it as satisfactory, it seems to provide for every important contingency that it was possible to foresee, and it promises a safety, permanence and stability that are too often wanting in similar organizations.

To enter upon a discussion of the personal credit due in the membership, the board of managers and of scientific directors, and in the Garden staff, would be an agreeable pleasure, but I must confine myself to the very earnestly made remark that the great success of the Garden has been due to the love of the institution and its work which has animated all concerned in it. It is this which has lent faithfulness, earnestness and energy and has incited to many acts of great sacrifice. If it could ever be said of any similar institution, we are able to say of this that it is a monument of loving service, in which work has been accepted in considerable part as its own reward. This is wholly true of Mrs. Britton's work in building up one of the most important departments of bryology in existence.

I dare not enter upon a detailed history of the Garden's development, and it has been so often and so recently recorded that I do not deem it necessary. An excellent account of its organization and of Columbia's relation to it, by Professor Underwood, can be found in the *Columbia Quarterly* 4: 278. 1903. Our charter was secured in 1891 and was amended in 1894. It was agreed upon that 250 acres of park lands should be set apart for our use and $500,000 appropriated for the museum building and conservatories, as soon as an endowment fund of $250,000 was obtained. This fund was completed in 1895, Columbia University making the first subscription of $25,000. With the election of Dr. N. L. Britton as Director-in-Chief, and his selection of a working staff, the preparations were complete and work began in 1896, the event which we are to-day celebrating. This was the year in which the first part of Britton and Brown's Illustrated Flora was published. Ground was broken for the Museum Building in December, 1897, and for the conservatories in 1898. The Museum was opened in 1899. In 1898 the bulk of the herbarium of Columbia College, numbering nearly half a million specimens, and of its botanical library, including more than 5,000 bound volumes, was turned over to the Garden, in trust and for its use, under certain stipulated conditions. Since then the herbarium has been more than doubled, and the library has been enlarged to 18,000 volumes. A vast amount of grading has been done, many miles of walks and roadways built, bridges erected, and a great increase in all the collections has been made.

Besides the *Bulletin* and the *Journal*, regularly published, the Garden has entered upon a work of a much more ambitious character. Utilizing the David Lydig fund, bequeathed by Judge Daly, it has begun the publication of an elaborate " North American Flora," the first parts of which have already been published. Provisions have been made also for the publication of colored plates of American plants.

Among the very important undertakings maintained have been extensive explorations, not only in the United States proper, but in such distant regions as the West Indies and the Philippines. A tropical station is maintained in Jamaica for the convenience of visiting botanists. At the Garden a scholarship fund is maintained, by which it is rendered possible for investigators desiring to pursue important studies here to be supported for a limited period.

A bird's-eye view only is permitted us of the botanical forces at present active in our city, including schools and classes, societies and botanical gardens and parks.

Botanical instruction, in the form of nature study, is now an integral part of our elementary school system, and is continued, in one form or another, in the higher grades. Spring and fall lecture courses and object teaching are conducted at this Garden for the benefit of the grammar schools of the Bronx, and it is to be hoped that provision may soon be made for extending the opportunity to the other schools of the City. Systematic instruction for the botanical training of teachers is given at the City Normal College, Teachers College, in the pedagogical department of New York University, and by the Brooklyn Institute of Arts and Sciences. Important work in the same direction, as well as in that of original research, is conducted at the summer school of science at Cold Spring Harbor. Columbia University provides ample and exceedingly varied botanical work in its different departments. Botanical teaching at the College of Pharmacy, now a department of Columbia University, dates back almost to the beginning of the College, in 1829. Although its work is technical, an effort has always been made to keep in sight its scientific basis.

At Columbia University itself, the department of botany is in charge of Professor Lucien M. Underwood, one of the most eminent, critical and conservative of botanical investigators, who has been accorded the status in universal botany that he merits. The bulk of the instruction work is under the immediate care of Dr. Carlton C. Curtis, and none better is given in any modern university. It seems most unfortunate that Dr. Curtis's great work should not be more generally known and more definitely recognized. This work is most ably supported by Professor Herbert M. Richards and Dr. Tracy E. Hazen in Barnard College, the department for women, which corresponds to Columbia College, for men. The instruction work at the New York Botanical Garden is of the most advanced character. Only those who have demonstrated their ability to pursue original investigations are admitted, and these are expected to engage while here in work of that character. More than half a hundred such pieces of original investigation have been conducted here in a single year.

Of local societies engaged in botanical work we have a number which are mere private associations, of a few persons, without formal organization, besides others to be mentioned. We have also a number, like the Linnaean Society, the Brooklyn Institute, the Staten Island Association of Arts and Sciences, the Bronx Society of Arts and Sciences, the West Side Natural History Society, and the local chapter of the Agassiz Association, which are engaged in the general pursuit of science, of which botany forms a part. Those devoted solely to botanical work of some sort are the New York Horticultural Society, which holds meetings, conducts lecture courses, and gives exhibitions, with the award of prizes; the Hulst Botanical Club of Brooklyn, a distinctly amateur organization; the Botanical Club of the Normal College, which aims to stimulate in its students and graduates a love of study, outside of that required by the regular course of instruction; and the Barnard Botanical Club, a somewhat similar organization, which aims to keep alive in the graduates a regard for the interests of the botanical department of that college, holds annually two regular meetings and provides one public lecture, and to which students of Barnard are eligible as members, after having performed one year of botanical work at the college.

Lastly, there is the Torrey Botanical Club, which endeavors to act as a central organization, representing in its membership that of all the other active botanical organizations in the city. Its present active membership numbers about 250, having increased 25 per cent. during the present year. It publishes three periodicals, holds two in-door meetings monthly, between October and May inclusive, and field meetings each Saturday during the season of plant growth. As has already been stated, an interest in plants from any point of view is the only botanical qualification required for membership, the nomination being made by some member of the Club and approved by the committee on admissions.

Among botanical gardens, it is not out of place for us mentally to include all the numerous and extensive horticultural establishments which abound in and about New York, among the stock of which is to be found such a great variety of plants of interest from botanical considerations. The public parks of this city are also to be justly regarded as affording important advantages for botanical work. Active and enthusiastic botanists are connected with them, and the planting, labelling and exhibiting are conducted with a view to interesting the public in the scientific basis of the work. The great collection of North American woods at the American Museum deserves special mention. People in this city who are interested in such subjects should also make themselves acquainted with the elaborate park system of Essex County, New Jersey, which has been laid out and organized with studious regard to future conditions and needs, and will undoubtedly develop important botanical features as time goes on.

Our own Botanical Garden you are to inspect to-day under unusually favorable circumstances. Even this, however, will give you but a very inadequate idea of the breadth and depth of its organization and character. There is scarcely a department of botanical work for the development of which provision is not made, the several departments being under the care of accomplished specialists. As you go about the grounds and enjoy the beautiful grades, the roads, walks, and bridges, you perhaps do not realize the immensity of the task involved in bringing them into existence and at the same time establishing and developing the scientific, cultural and educational departments. From the time of its foundation, the Garden has had more than one interest clamoring loudly for the expenditure of every available dollar. Its economical and efficient management has usually contrived to divide that dollar and make each part of it do the work of the whole.

In the conduct of any growing enterprise not only does each step taken become a new point of departure, but new centers of work become established by the division of the old; and so our review would not be complete without a glance at the most important requirements for the future. One of these is the organization of a well-equipped botanical department at New York University. One of the leading universities of the country, with well-organized departments and many hundreds of students, it seems a continued misfortune that it should not be in a position to utilize the many facilities which we have to-day considered, and equally so that our science should not profit by the stimulus and support which would result from the maintenance of an adequate center of activity at University Heights.

Our Botanical Garden suffers greatly from the want of a larger endowment fund. Its charter provides for the construction and maintenance of its framework, but back of this lies the necessity for supporting its higher life, and for this support we must naturally look to its endowment. The two should keep close pace. The crown of the greater tree demands a greater root system for its support. Our plant has increased wonderfully in ten years, both in size and in the intensity of its activity, while the endowment has remained stationary. Its increase to the sum of $1,000,000 has been undertaken, and the amount is none too large and can come none too quickly. One of the special needs of the Garden, or rather of this part of the country through its Garden, is a department of forestry. From an economic point of view, this is by far the most important department of botany at the present time. Our need of increased forest resources is already alarming to every serious political economist. When an attempt is made to provide them, we find that we do not know how ; that every tree must be known separately, and that until this is done practical operations must fail ; and that the acquisition of this necessary knowledge is as slow as the growth of the trees themselves. It is urgently necessary that such centers of investigation should be established in numbers. Scarcely anywhere is there an institution that combines so many advantages for a successful organization of this kind as here. Our Club has this year undertaken to arouse interest in the subject by providing a course of ten field lessons, conducted by competent instructors, and open to all our members, without charge.

Did time permit, I should be glad to speak on this occasion of the special needs of our Club. In a general way we should get back to the work for which we were originally organized — the study of our local flora, at present construed as that within a 100-mile radius of this city. To do it properly provides ample work for years to come. It is a work of important scientific value, yet includes popular features calculated to interest every member. All that is needed is a leader, and this is the point of difficulty. He must be a capable botanist, and he must give practically his whole time to the work. This means that he must be compensated, and this is possible only through an endowment fund, or through a very large membership list, for both of which we earnestly hope. If 200 others of the 10,000 or more persons of this section whose interest in plants entitles them to become members of the Club would do so, there would be ample provision for the undertaking of this work.

EARLY BOTANICAL ACTIVITY
IN THE DISTRICT OF COLUMBIA

Frederick V. Coville

EARLY BOTANICAL ACTIVITY IN THE DISTRICT OF COLUMBIA.

By FREDERICK V. COVILLE.

(Read before the Society May 6, 1901.)

Early in the winter I received from Mr. W. B. Bryan an invitation to prepare a paper on early botanical activity in the District of Columbia. Knowing that Professor Lester F. Ward had for some years been interested in this subject, I suggested to him that he assume the task. Professor Ward felt, however, that he had not time to treat the subject to his own satisfaction and he offered to turn over to me the material he had accumulated. To this courtesy is due chiefly whatever of value this paper may contain.

The documents received from Professor Ward were as follows:

A Description of the District of Columbia. By D. B. Warden. 1816.

Proceedings, Constitution, etc., of the Washington Botanical Society. Bound manuscript.

Florula Columbiensis. [Anonymous.] 1819.

The Washington Guide. By William Elliot. Edition of 1837.

Florae Columbianae Prodromus. By John A. Brereton. 1830.*

---

* In addition to the documents listed, there was in Professor Ward's possession a botanical work from the library of the Geological Survey, which, while not bearing on the botanical history of the District of Columbia, has a certain local interest from the fact that it was published in Georgetown, D. C., in 1814. It was " printed by J. M. Carter "

Turning to the general conditions that existed in Washington after the establishment of the seat of government in this city in 1800, one would expect to find the government itself conducting some sort of botanical investigation. But no national museum existed at that time, a botanical garden although planned by George Washington had not been established, nor had the first of the geographical surveys and explorations been begun. Thomas Jefferson's well-known interest in the science of botany apparently was never brought to bear upon the District. Indeed Jefferson's appreciation of this branch of scientific research served in one instance to accentuate the poverty of the government's equipment for it. That occasion was the Lewis and Clark expedition across the continent to the mouth of the Columbia River and back, in 1804 to 1806. In his instructions to the commanders, Jefferson said: "Other objects worthy of notice will be: The soil and face of the country, its growth and vegetable productions, especially those not of the United States; .... Climate as characterized by the thermometer; .... the dates at which particular plants put forth or lose their flower or leaf." The collection made by Lewis in compliance with these instructions was not kept at Washington, for there was no government botanist, but was sent to Philadelphia, then the center of botanical research in the United States, and placed in the hands of Benjamin S. Barton, a prominent botanist of the time.

Likewise neither the horticultural nor the agricultural

and is entitled, " A synopsis of the genera of American plants, according to the latest improvements on the Linnæan system: with the new genera of Michaux and others. Intended for the use of students in botany." The name of the author does not appear in the book, but on page 7 of Brereton's Florae Columbianae Prodromus, described later in this paper, the authorship is credited to O. Rich.

12

interests of the community seem to have had any effective influence on the development of botanical science. In the past few decades both horticulture and agriculture have derived enormous benefits from botanical research, which in return has received from these arts handsome financial support. But the usefulness of the intimate mutual relations that exist to-day was not at that time recognized, and the Columbian Agricultural Society, organized in 1810, dealt exclusively with the encouragement of what was then considered more practical agricultural development. The nurserymen of those days, too, more than the nurserymen of to-day failed to develop the scientific side of their art. Mr. Thomas Maine, a Scotch gardener who settled near Georgetown about 1804, appears to have been the first nurseryman of the District.

From what has been said it is evident that several of the circumstances which might have been expected to incite botanical activity in the District had no such influence. What follows will show that the two main influences which actually awakened it were, first, the interest of cultivated men and women, and, secondly, the special interest of members of the medical profession, who by their training and their practice were brought into an intimate relation with botanical science. The same influences were everywhere manifest in the early development and progress of botanical science throughout the United States.

The earliest of the documents mentioned at the beginning of this paper was a book of 212 pages, published in Paris in 1816, and entitled "A chorographical and statistical description of the District of Columbia, the seat of the general government of the United States, with an engraved plan of the District, and view of the Capitol." Its author, D. B. Warden, was an American who was at the time residing in Paris. In dedicating the book to Mrs. Custis (Mrs. Elizabeth Park Custis, the granddaughter of Martha Washington), the author says:

"It may not have escaped your recollection, that you kindly honored me with your advice to occupy my leisure hours, at Washington, in examining the interesting objects of that magnificent situation. You were even pleased to accompany me in some of my excursions, and to honor me by an introduction to your relations and friends, to whom I feel grateful for valuable and unwearied attentions. I brought to Paris my notes and collections of plants, minerals, and insects, which I had not leisure to examine as long as I exercised my public functions. My labors, however, have been of little avail in struggling against fortune, to whose capricious empire I have been forced to submit.

"Since the suspension of my consular powers, I have occupied myself chiefly with subjects relating to the United States; and I have prepared for your acceptance this sketch of the District of Columbia, which I flatter myself will serve to perpetuate the fond remembrance of the friendship and confidence with which you have so long honored me."

This extract sufficiently explains the circumstances under which the book was written, and to some extent makes clear the conditions under which the materials for it were collected. It is full of concise and interesting information about the Washington of that day, ranging from statistics of the commerce of the upper Potomac and the price and cost of maintenance of slaves, to a description of the characteristics of the "women of Columbia," a statement of the amount of the fine for "profane swearing and drunkenness," which was 83 cents, and an account of the best method of preventing hogs from passing underneath hedges. The part with which we are now most concerned, however, is pages 191 to 209, entitled, "Florula Colum-

biana, sive enumeratio plantarum in territorio Colum-
biae sponte nascentium; or, catalogue of the plants,
shrubs, and trees which grow spontaneously in the District of Columbia.'' One hundred and seven genera
and 134 species are enumerated, with both technical and
popular names and statements of their uses together
with other items of interest. In a prefatory paragraph
the author says:

"This catalogue contains only the specimens which I was
able to collect in a few excursions through this district, and
consequently is far from being complete. The collection,
however, has acquired value, from being examined by the celebrated botanist, Correa de Serra,* to whose generous friendship on this, as on other more important occasions, I feel
deeply indebted."

A critical review of the contents of the catalogue is
more suitable for a botanical than a historical society,
and we may therefore pass the subject by with the statement that though the list of species is small, containing
less than one-tenth the number now known in the District, it is presented in excellent form, and is conclusive
evidence of an intelligent interest in botanical science
among the educated people of Washington at that early
period.

Professor Ward's second document, a book containing the original "Proceedings, Constitution, etc., of the
Washington Botanical Society," in manuscript, contains much of historical interest. The book, it should
be recorded, was presented to Professor Ward about
1881 or 1882 by Mr. James Anglim, the predeces-

* Joseph Correa de Serra, the minister of Portugal to the United
States, took an active interest in the promotion of the natural sciences in
America, and among other worthy acts assisted in procuring the means
for Thomas Nuttall's botanical exploration of Arkansas Territory.
Nuttall dedicated to him his Genera of North American Plants, pub-
lished in 1818.

sor of the book firm of W. H. Lowdermilk & Co., who received it among some second-hand books that had come into his possession.

The opening entries in the proceedings are as follows:

" Thursday, March 13, 1817.

" A meeting was held this day at Davis's Tavern, pursuant to public notice for the purpose of forming a Botanical Society—Doctor John A. Brereton being called to the chair and John Underwood appointed secretary; it was on motion

" *Resolved,* That a committee of three be appointed to draught a constitution for the Society above mentioned, and that they report the same at the next meeting.

" *Resolved,* That the Rev⁴ Doctor James Laurie, George Watterston, Esquire, and Doctor Alexander McWilliams compose the committee.

" The secretary was instructed to inform those gentlemen of their appointment and the meeting was adjourned to Thursday, the 20th inst.

" Thursday, March 20, 1817.

" The meeting was held at Davis's this day pursuant to adjournment.

" The committee appointed at the last meeting reported a Constitution for the Society, which after having been discussed and amended was unanimously adopted."

Probably no better idea of the aims of the Society can be secured than from a reading of this Constitution, which was as follows:

CONSTITUTION
OF THE
WASHINGTON BOTANICAL SOCIETY.

" ARTICLE 1.

" This association shall be denominated the Washington Botanical Society.

13

" 2.

" The Society shall be composed of honorary and resident members.

" 3.

" The objects of this association shall be to collect, arrange, preserve and describe all the vegetable productions within the limits of the District of Columbia whether indigenous or exotic and to detail when practicable all their medicinal, esculent and other properties.

" 4.

" To publish quarterly, if deemed necessary, whenever the Society shall have obtained a full knowledge of all the vegetable productions of the said District a Flora, with colored plates; each plant to be classed and arranged according to the Linnean system and described if known under the direction of the president and vice-president of the Society.

" 5.

" The officers of the Society shall consist of one president, two vice-presidents, a secretary, treasurer and three curators, who shall be chosen by ballot on the first Monday in March annually.

" 6.

" There shall be a corresponding committee consisting of the president, vice-president and secretary of the Society, whose duty it shall be to correspond with botanists and other persons and to conduct all correspondence whatever in relation to the objects of this Society.

" 7.

" No official situation in this Society shall exempt any one from performing duties in common with other members.

" 8.

" The officers of the Society shall be chosen from the resident members and shall be elected by a majority present at the stated meetings.

" 9.

" Five members, including officers, shall form a quorum to transact business, except to alter the constitution and elect honorary members; in which cases two-thirds of the members, including officers, shall be required to form a quorum.

" 10.

" The election of new members shall be by ballot and by a majority present at the stated meetings of the Society.

" 11.

" All resident members shall pay into the hands of the treasurer every year [a sum not exceeding] five dollars, which together with such other moneys as may be collected, shall be distributed as the Society may direct.

" 12.

" The Society shall assemble on the first and third Mondays in every month during the spring, summer and fall and once during the winter season.

" 13.

" Each member shall pay a fine of one dollar for every stated meeting and fifty cents for every special meeting he neglects to attend; and no excuse shall be received except such as may be deemed reasonable by the Society.

" 14.

" The Society shall be divided into four committees, each committee to consist of the one-fourth of the members including officers whose duty it shall be to collect and preserve the specimens of plants they may find within the portions of the District assigned them for examination, which specimens shall be preserved in a Herbarium prepared for that purpose by the Society and placed under the charge of the curators.

" 15.

" Every committee shall be furnished with a herbarium to preserve the duplicate plants they may procure in the course

of their researches, and when the class, order, genus and
species of a plant cannot by them be ascertained, it shall be
their duty to lay it before the Society at their stated or
special meeting to be examined and arranged.

" 16.

" It shall be the duty of the president to preside at each
meeting to preserve order, to call special meetings, to super-
intend the concerns of the Society and in all equal divisions
to have a casting vote.   It shall also be his duty to apportion
the District into four sections and to assign to each committee
a separate section.

" 17.

" The senior vice-president present shall preside during
the absence of the president.

" 18.

" The secretary shall keep a minute of the proceedings of
each meeting of the Society, note the members present and
carefully transcribe in a book to be provided for that purpose the class, order, genus, species and description of the
plants obtained by the Society; he shall also give due notice
of the stated and special meetings of the Society when necessary.

" 19.

" The treasurer shall collect all moneys due to the Society
and discharge all bills accepted by the president.   He shall
keep regular accounts of the receipts and expenditures, subject to the examination and inspection of any member that
may require it.

" 20.

" The curators shall take charge of the Herbarium, of all
communications and donations to the Society, which they
shall arrange under their respective heads, and also of all
specimens presented that it may not be proper to place in
the herbarium together with all drawings, books, etc., belonging to the Society and keep lists of the donations with the
names of the donors and their places of residence."

The signers of this constitution were:

W. A. Bradley,	S. Eliot, Jr.,
J. M. Moore,	Geo. Watterston,
J. W. Hand,	John Boyle,
James Laurie,	James Kearney,
Henry Huntt,	Jno. Underwood,
Wm. Elliot,	J. A. Brereton,

Alex^r McWilliams.

The officers of the Society for the first year were as follows:

Revd Doct^r James Laurie,	President.
Saml Eliot, Jr.,	1st V. President.
	2d do. vacant.
George Watterston,	Secretary.
John Boyle,	Treasurer.

Doctrs. Alex. McWilliams,
    John A. Brereton, } Curators.
    Henry Huntt,

Revd Doctor James Laurie,
Samuel Eliot, Jr., } Corresponding Committee.
George Watterston.

At this distance the exacting character of the obligations imposed on the members of the first Washington Botanical Society, such as the fine of one dollar for non-attendance at meetings, may furnish some amusement, but the constitution as a whole is good evidence of the high aims of the Society's founders.

At the meeting of April 7, 1817, after the election of three additional members, the acting president approved the following committees for the examination of the four sections of the District flora:

" From the Tiber to the President's House, North and South," Messrs. Boyle, Brereton, Underwood and Hand.

" West of the President's House and North of the Penn-sylvania Avenue to Rock Creek,'' Messrs. Kearney, Moore and Bradley.

" From the Tiber East and North to the District line,'' Messrs. McWilliams, Watterston, S. Eliot and Thompson.

" Beyond Rock Creek to the District line,'' Messrs. Laurie, Huntt and Steiner.

With such a beginning the Society entered upon an active existence, holding frequent meetings, at which the plants collected by the members were presented, examined, and discussed.

In addition to the original members the following residents of the District were elected:*

—— Franzoni,   ——— Steiner,
Joseph Milligan,   Rod. Schaer,
James Thompson,   Nicholas Worthington,
                Ann Davis.

The election of the member last named was attended with considerable formality, as thus recorded in the minutes of the meeting for April 14, 1817:

" On motion, a committee consisting of Messrs. Bradley, Brereton and Watterston, was appointed to wait on Miss Ann Davis, and offer her in the name of the Society, a member-ship of the same. The committee withdrew, and in a short time returned, and made report that they had waited on Miss Ann Davis, agreeably to the resolution, and were directed to inform the Society that she cheerfully accepted the honor offered her, but hoped that the necessity of her attendance at the meetings would be dispensed with; to which the Society consented.''

On May 19, 1817, the Society adopted a series of by-laws relating chiefly to the methods of bringing in and examining plants.

* The following American botanists were also elected members of the Society: Jacob Bigelow, of Boston; William Darlington, of New Jersey; and W. P. C. Barton, of Philadelphia.

In the year 1817 twenty-eight meetings were held; in 1818, twenty-six meetings; in 1819, twenty-one; and in 1820, fifteen.

During this time the Society prosecuted its objects faithfully and it appears enthusiastically. Among items of interest may be cited the following:

On July 28, 1817, a resolution was adopted

"That the Corresponding Committee address a circular to the men of science of the District and the adjoining states soliciting such information as they may possess in relation to the medicinal, esculent, poisonous and other properties of such native plants as they may be acquainted with."

On September 12, 1817,

"The following resolution was offered by Mr. Watterston, which was ordered to lie on the table:

"Resolved, That a committee be appointed to present a petition to Congress requesting the passage of a law authorizing a lottery for the purpose of establishing a Botanical Garden in the City of Washington under the superintendence of this Society."

On November 7, 1817, it was resolved

"That the Washington Botanical Society consent to become members of the Columbian Institute provided they so alter the constitution as to admit said Society into the Committee on Botany and Agriculture."

This proposition was not carried out.

On March 27, 1820, the following resolution was presented by Mr. Watterston:

"Resolved, That with a view to form a National Herbarium it shall be the duty of each member to preserve two or more of such specimens as he may collect or procure, to be submitted to the Society at their stated meetings, and a selection of the best shall be made and transferred to the General Herbarium under the care of the curators, whose business it shall be to arrange them at the close of every year, name them, and deposit the said collection in the Herbarium of the United States."

At the first meeting in the year 1821 it was

"Resolved, That the meetings of this Society be hereafter held at the building where the Genl. Post Office is kept & in the room lately occupied as the Library of Congress.

"Resolved, That a committee of two be appointed to wait on D. Thornton & obtain his assent to our having the use of said room and to make all arrangements which may be necessary for removing thither all the books and other property of said Society."

That interest in the Society had begun to wane was evident from the fact that only five meetings were held in 1821, and that an unhealthy condition of affairs existed was further evidenced by an entry in the minutes of the first of the seven meetings in 1822:

"On motion it was resolved to examine the state of the funds of the Society. On examination it was found that the Society was indebted to the treasurer [Mr. Boyle] 17 dollars and 84 cents, to Mr. Brereton 5 dollars and to William Elliot 1 dollar."

At the meeting of May 6, 1822, in response to a motion, the secretary stated that the following were the only members of the Society, the others being excluded by the by-laws for non-attendance:

Dr. Brereton,	Mr. Underwood,
Dr. McWilliams,	Mr. Elliot,
Mr. Boyle,	Maj. Kearney.

Only a single meeting was held in each of the years 1823, 1824 and 1825. In 1826 a meeting was called for March 20th, and adjourned to the 27th. On this date the Society, apparently with a consciousness of impending death, passed the following resolutions:

"That it be made the duty of the president of the Society to collect all the books belonging to the same, to have them deposited in a case in the Washington Library, under the charge of the librarian; that each member may have access to the same agreeable to the rules of the Society."

"That the president, Dr. McWilliams, be authorized to take charge of the Herbarium till further order be taken."

The Society then adjourned sine die. It is doubtful whether the going out of any scientific association ever took place in a more orderly manner than that of the Washington Botanical Society in 1826.

Although the Society itself was dead, it left, either directly or indirectly, certain published records of its work on the flora of the District. The first of these bore on its title page the following:

"Florula Columbiensis: or a list of plants found in the District of Columbia; arranged according to the Linnæan system, under their respective classes and orders, &c., and exhibiting their generally received common names, and time of flowering, during the years 1817 and 1818. Washington: printed for the Washington Botanical Society by Jacob Gideon, Jun., 1819."

The work is a 14-page pamphlet giving a bare list of the technical and popular names of 296 species of flowering plants with the date of their observation in 1817 and 1818. It apparently was intended as a working list for the members of the Society.

The second publication based on the work of the Society appeared as a chapter of the first edition of William Elliot's "Washington Guide," published in 1822. It is headed "Botany of the District of Columbia." The introductory portion is as follows:

"To render our work as scientific as its compass will admit of, we have prevailed upon a member of the Washington Botanical Society to give us a brief sketch of the botany of the District. He has politely· furnished the following, arranged after the Linnean classification. We regret, however, that room cannot be found in this small volume to comprise the objects to which he alludes in the following letter accompanying the sketch:

" WASHINGTON, October 12th, 1822.
" Mr. Elliot,

I herewith send you, agreeable to promise, a list of the plants growing within the District, which have as yet been noticed. It was my intention to give you with each genus, the time and duration of flowering from 1817 to 1822, inclusive, together with the localities, soil, exposure, &c., and to have added the natural orders of Linnæus as well as the more fashionable one of Jussieu; and, also, to have added some remarks on a few of the species enumerated, that have been doubted as growing immediately within this District: but from the space allowed I shall wait a more favorable opportunity.

"Very respectfully,
" Your obedient servant,
" J. A. BRERETON."

After this introductory note follows a list of the plants of the District, 458 in number, under the heading "Florula Columbiana." In subsequent editions of the "Guide" the same list was repeated.

In the year 1830 appeared Brereton's " Florae Columbianae Prodromus," the full title-page of which is as follows:

Florae Columbianae prodromus, exhibens enumerationem Plantarum quae hactenus exploratae sunt: or a prodromus of the flora Columbiana, exhibiting a list of all the plants which have as yet been collected. Compiled by John A. Brereton, M.D., U. S. Army. Washington, printed by Jonathan Elliot, and sold at his store on Pennsylvania Avenue. 1830.

The work is an 86-page 16mo, containing a list of 860 species of plants, arranged according to the Linnean system. Seldom is more than a single line devoted to a species. In the preface the author said:

"During the spring of the year 1825, after the dissolution of the late Washington Botanical Society, a few gentlemen of this city, devoted to the science of Botany, formed an association, with an earnest determination to explore and to investigate, *de novo*, the indigenous plants growing in the District of Columbia. The association, under the name of "The Botanic club," consisted of Wm. Mechlin, Wm. Rich, Alex'r McWilliams, M.D., the compiler, and during the following year, of James W. Robbins, M.D., but who, afterwards, in the succeeding one, removed from the District. The Prodromus, herewith submitted to the public, is, so far, the result of their investigations to the present time; and, there is no doubt that their contemplated Flora will contain a much more augmented catalogue, more particularly among the *species,* than is now presented."

It is desirable to introduce here an interesting document only very lately brought to light. A few days ago my attention was called by Mr. W. B. Bryan to the fact that the George Watterston papers recently acquired by the Library of Congress contained some memoranda relating to the Washington Botanical Society, of which, the journal of the Society shows, Mr. Watterston was for the first three or four years an active member, and for the first two years the secretary. From a rather hurried examination of the Watterston papers, which I was enabled to make through the courtesy of Mr. C. H. Lincoln, chief of the manuscript department, a single document was found bearing upon the present subject. It is the original draft of what was apparently a newspaper communication by Mr. Watterston after the issue of Brereton's Florae Columbianae Prodromus in 1830. It is as follows:

" Error Corrected.

" In the year 1816 there came to Washington a Scotchman named Whitlaw who gave notice that he would lecture on botany. He possessed some splendid transparencies of the fructification and other parts of plants executed by J. Thornton of London, where he had obtained them, and which served admirably for illustrations of this beautiful science. A class of citizens was formed who attended his course of lectures such as they were. They were delivered extemporaneously and though in the broad accent of the Scotch and in no very elegant language, his explanations, aided by the fine transparencies he possessed, were sufficient to give his class a very correct idea of the elements of botany. After he had finished his course, it was suggested by George Watterston, one of the gentlemen who had attended his lectures, in order that the knowledge of the science which had thus been acquired might not be lost and that it might be improved and applied to some useful purpose, it would be proper to form a society, the object of which should be to explore the District of Columbia, then embracing an area of ten miles square, and ascertain its botany. The society was accordingly established and consisted of the following members, viz.: Messrs. J. Boyle, J. Kearney, W. A. Bradley, Dr. Laurie, Geo. Watterston, Dr. A. McWilliams, and Dr. Brereton, all with the exception of the *two last*, members of Whitlaw's class.

" Committees were appointed for the different parts of the District, the members of which weekly brought to the place of meeting, a room in Dr. Brereton's house on the Penna. Avenue, the collections of plants they had made in their various excursions and after being examined by the whole Society and ascertained were recorded in its journal by the secretary, G. Watterston. Dr. Brereton, who was lame and unable to accompany any of the committees, remained at home and took charge of the specimens as they were brought in and would occasionally employ himself in ascertaining their names from the botanical works in his possession. These books were considerably increased from time to time by the annual pecuniary contributions of the members; but what became of them after the dissolution of the society is not known. The Association continued in existence till specimens of almost all the plants of the District had been collected, named, and recorded in the journal which was left in the possession of Dr. Brereton. Soon after the removal of this gentleman to a house near the War Department, the Society having accomplished in a great degree the object of its organization, ceased to exist and was virtually dissolved. Some years after its dissolution a work appeared purporting to be the production, the result of the labors of Drs. Brereton, McWilliams, Mechlin and Rich, etc., containing the catalogue of the plants collected by the Botanical Society and which contained very few if any more than had been named and recorded in the journal of that Society. And yet its labors are scarcely mentioned and the whole work spurned by the persons named above. Dr. Brereton flourishes as the compiler and principal agent. I have deemed it my duty to strip the daw of his borrowed feathers and to point out the real source from which this catalogue was derived in justice to them and for the sake of truth.

"The journal of the Society must still exist and is no doubt in the possession of the family of Brereton. A reference to this journal will establish the truth of what has been stated. Brereton's only merit seems to be that of having arranged the plants according to their classes and order, their names having been entered and classed promiscuously in the journal. I have thus deemed it a duty to point out the real source from which the Florae Columbianae Prodromus was derived in justice to the members of the Botanical Society and for the sake of truth.

"G[EORGE] W[ATTERSTON]."

This document is chiefly interesting as showing the circumstances that led to the formation of the Washington Botanical Society. Regarding the point in controversy between Mr. Watterston and Dr. Brereton, it may be said that the journal of the Society, according to a compilation made by Professor Ward, contained the names of 370 plants, while the Prodromus, as already stated, enumerated 860 plants. All misunderstanding would doubtless have been avoided if the author of the Prodromus had credited the Society with a contributing share in the make-up of the list.

The more pretentious "Flora" contemplated by the constitution of the Botanical Society and later by Dr. Brereton and his associates never was issued, and Dr. Brereton's Prodromus remained for nearly fifty years the standard list of plants of the District of Columbia. It may fairly be considered the principal published record of the activities inaugurated by the Washington Botanical Society in 1817, the first scientific society of the District of Columbia.

SOUTHERN BOTANISTS

F[rank] Lamson-Scribner

Southern Botanists.*

BY F. LAMSON-SCRIBNER.

Ladies and Gentlemen: At this season (March 26th) there may be found in our rich woodlands, brightening the deep shades with its pure white flowers, a little plant known to botanists the world over as *Jeffersonia.*

A more delicate tribute cannot be paid a worker in the science of botany than to name for him some new or undescribed plant. Only those who by their direct labors have increased our knowledge of the vegetable kingdom, or who have in some way materially aided in the advancement of the science, have been thus honored.

It is the purpose of this lecture to pass in review some of those who have been thus distinguished in the annals of American botany, or who have thus become identified with the plants of the Southern States. The number is far too great for all to be included in a popular lecture, nor can I claim your time to say all that I would like to say of those even who are ranked most prominent.

Jeffersonia diphylla commemorates one whose name and fame are so well known to all that I scarcely need more than mention him. Although Jefferson was not a botanist, his recognition of the science and the successsul encouragement which he gave to its prosecution will ever be remembered by botanists in the plant which bears his name. In the development of our nation and national politics this name will ever stand among the first and foremost, and in the development of natural history in America it holds a no less prominent position. "It is probable," says G. Brown Goode, "that no two men have done so much for science in this country' as Jefferson and Agassiz; not so much by their direct contributions to knowledge as by the immense weight they gave to scientific interests by their advocacy."†

To Jefferson's interest and influence was due the organization of the first government exploring expedition, that of Lewis and

---

* A lecture delivered at the University of Tennessee, March 26, 1889.

† "The Beginnings of Natural History in America." Liberal use has been made of this address by Prof. Goode in the preparation of this lecture.   F. L. S.

Clark, into the far Northwest—the precursor of all like enter-prises carried on by the general government, culminating in the present magnificent Geological Survey,

A little incident illustrating Jefferson's scientific enthusiasm is thus related. On going to Philadelphia to be inaugurated Vice-President, he carried with him a collection of fossil bones which he had obtained in Green Brier county, Virginia, together with a paper in which were formulated the results of his studies upon them. This was published in the Transactions of the American Philosophical Society, and the animal which the bones illustrated is still known as Megalonyx Jeffersoni. The spectacle of an American statesman coming to take part as a central figure in the greatest political ceremony of our country, bringing with him a lot of bones and an original contribution to science is one long to be remembered, and is not likely soon to be repeated. Botanists have done well to preserve in their special field the memory of this man, remarkable as he was great.

A tropical American genus of woody climbers, Banisteria of Linnæus, was dedicated to John Bannister, who settled in Virginia some time prior to 1668, and who published in 1686 a "Catalogus Plantarum in Virginia Observatarum," which was the first system-atic paper upon natural history emanating from America. He was an artist, for with his notes and dried specimens transmitted to Bishop Compton and John Ray, of England, he sent drawings of the rare species which he found. Ray says of Bannister in his "Historia Plantarum," "erudissimus vir et consummatissimus Botanicus."

The memory of John Bannister is still cherished in Virginia where his descendants are numerous.

That attractive little spring flower *Claytonia Virginiana*, which enjoys the honor of having been the first to fall into the hands of Dr. Gray, when a student, to be analyzed by him, and which with one or two others disputes the right to be recognized as the " May Flower" of our ancestors, keeps fresh in our memories one of the earliest devotees of botanical science in America, John Clayton, of Virginia. For fifty years Clayton was clerk of Gloucester county and during all this period he spent a great deal of time in exploring the region about him and in describing the plants which he found.

He was a correspondent of Gronovius and also of the great Linnæus, both of whom afforded him much aid in his botanical pursuits. Clayton's " Flora 'Virginiana," the first of its kind in this country, began to appear in 1739, subsequent portions being published in 1743 and 1762. At the time of his death he left two volumes of manuscripts, and an herbarium with marginal notes and references for the engraver who should prepare the plates for his proposed work. All this material, the result of many years of labor, was destroyed by fire during the Revolutionary War, and thus perished what was probably one of the most important works on American botany written before the days of Gray and Torrey.

The author of the first work written in America on the principles of science (Botany and Zoölogy), was, like Clayton, a resident of Virginia. The low-creeping, evergreen vine, known to every one as Partridge-berry, was named by Linnæus in honor of this author, Mitchella repens. Dr. John Mitchell was born in England, but he early came to Virginia where he spent nearly fifty years practicing medicine and promoting science. He was a man of broad culture and was one of the earliest chemists and physicists in America. His political and botanical writings were numerous and were always well received, and Mitchell's map of North America is still an authority in boundary matters. It has been said of Mitchell and Clayton that together they gave to the botany of Virginia a distinguished lustre.

Mark Catesbey spent a dozen years in Virginia and the Carolinas, from 1712 to 1725, collecting and making paintings of birds and plants, and his magnificent, illustrated work on the Natural History of Carolina, Florida and the Bahamas, is still of very great value to students. A Rubiaceous genus of the Antilles perpetuates his name.

John Bartram, whose home was near Philadelphia, but who extended his botanical explorations into some of the Southern States, was styled by Linnæus "the greatest natural botanist in the world." He was made botanist to His Majesty for the Floridas by George III., and was given a pension of 50 pounds a year. He was a collector rather than an investigator, and did botany great service by supplying plants and seeds to Linnæus and other European botanists. He is best remembered in the so-called Bartram oak, Quercus heterophylla, and in the unique botanic garden which he established on the banks of the Schuylkill, near Philadelphia.

An herbaceous plant of the Old World, Michauxia, commemorates the author of the first "Flora of North America," and the most untiring explorer this country has ever seen. Although André Michaux the elder cannot be ranked as a Southern botanist, as he spent only a few years in the Southern States, he so identified himself with our flora, and being the first botanist to cross the mountains, in 1795, from the East into Tennessee and Kentucky, that we cannot pass him wholly unmentioned.

Of the 1700 species of plants enumerated in Gattinger's "Tennessee Flora," over 135 were originally described and named by Michaux, and many of these were found by him, some for the first time, within our State limits. Michaux, with his son's assistance did more than any other man to diffuse a knowledge of our forest trees, particularly the oaks. His "Flora" was published in 1803.

Passing over a number of names scarcely less noteworthy than those already mentioned, we come to those who have been most active in the investigation of Southern plants within the present century.

A yellow-flowered composite plant of the sand hills of Georgia and Florida, named by Nuttall, *Baldwinia multiflora*, calls up before us one of our pioneers in the field of botany, Dr. William Baldwin. Dr. Baldwin was born in Pennsylvania in 1779, educated at the University of that State, and in 1811 removed to Georgia. He was a man much beloved by his associates, of whom Stephen Elliott was one, and was possessed with a most amiable character. He studied very minutely the difficult family of sedges, and the yet more difficult genera of grasses, *Paspalum* and *Panicum*. One of his best botanical papers was published in the Trans. Am. Phil. Soc., of Phila., and another in the American Journal of Science, the first relating to sedges, the second to grasses. He was appointed botanist to the expedition under Maj. Long, but his health, never very strong, failed completely during the journey to the field, and he died at Franklin, Mo., in the forty-first year of his age. His collections of Southern plants are now in the Herbarium of the Acd. Nat. Sci., of Phila., where they are guarded with care and justly prized.

Muhlenberg, in Nuttall's "Genera Plantarum," did honor to one of the most distinguished of our early botanists by naming for Stephen Elliott, a shrub of the Heath family, which grows in the dry, rich soils of Southern Georgia, *Elliottia racemosa*. Elliott was born at Beaufort, S. C., in 1771. He was educated at Yale, and while a senior in college he was spoken of as being possessed of more science and general information than was often found in one of his age and standing. He graduated with one of the highest honors, and returning home applied himself to agricultural pursuits. The people of his State, recognizing his marked ability, elected him first to the State Legislature, and then to the Senate. While a member of the latter body he took a leading part in all important business, and was the originator of the "free school system" of South Carolina, and the Bank Bill creating the "Bank of the State."

He was for a time President of the South Carolina College and later Professor of Botany and Natural History in the Charleston Medical College. He was the first, and during his lifetime the only President of the Philosophical Society of Charleston. The versatility and vigor of Mr. Elliott's mind may be seen in the variety of attainments in which he excelled. Beginning his career as a legislator, in which capacity he served for many years, he took prominent and leading parts in many of the important measures of his day. And it was while engaged in public and in engrossing financial business that he found time for literary and scientific pursuits, which alone would have placed him in the foremost rank among men of letters.

Botanists remember and esteem Elliott for his grand work entitled a "Sketch of the Botany of South Carolina and Georgia," published in two volumes in 1821 and 1824. The technical descriptions of the species enumerated are given in both Latin and English, and these in each case are followed by a more extended account in English in which are given the habitat, time of flowering, local names and often the reputed medicinal properties and other points of historical interest. The whole work, embracing

1 349 pages, together with a dozen finely executed plates illustrating 48 species of sedges and grasses, exhibits great scientific accuracy and an unusual amount of care in its preparation. Such was the work which the author modestly styles a "sketch." It was the chief authority among botanists for this latitude prior to the appearance of Chapman's "Flora of the Southern States," in 1860, and to-day it is one of the works which all working botanists feel that they must possess or be able to consult. Until one has written a book, and especially a book where almost every line is the statement of a fact learned, for the most part, from original observation, can they appreciate the amount of patience and labor involved in the preparation of such a work as Elliott's "Sketch of the Botany of South Carolina and Georgia."

In the preface to the second volume, Elliott, in acknowledging his indebtedness to others for assistance, briefly refers to several botanists who come within the scope of this lecture, and I cannot do better than to quote Elliott's own words in speaking of them. "To those who have aided in collecting the plants from which this sketch has been compiled, the author feels his manifold obligations; he wishes to express them particularly to Mr. James Jackson, of Louisville, Ga., from whom he has received many new and many rare plants and whose notes have always rendered his specimens more valuable. To Samuel Boykin, of Milledgeville, who, residing in a most interesting district of country, has added much to the author's knowledge of its flora by the valuable collections of specimens occasionally sent him.

"To Mr. N. Herbemont, of Columbia, S. C., for many specimens of rare plants collected around Columbia and in the upper districts of Carolina.

"To Dr. William Baldwin, of the United States Navy, a botanists of distinguished talents and indefatigable activity, who while residing in the southern districts of Georgia, communicated many new species published in the earlier portions of the "Sketch."

"But principally to the late Dr. James McBride, a tribute is due, not only for the services which he himself actually rendered, but for the contributions which he induced others to offer. Devotedly attached to science, he had the talent to make it popular wherever his influence extended. Profoundly skilled in his pro-

fession and high in the confidence of his fellow citizens, he fell a
victim to the fatigues and exposure of an extensive practice. In
the midst of a brilliant career, with prospects of increasing usefulness and extended reputation, he died at the early age of 33."

In dedicating to this gentleman the genus, *Macbridea*, Elliott
says, " I have named this genus in commemoration of the late Dr.
James McBride, whose untimely death, medicine and natural history and an admiring country equally deplore."

There are a number of plants in our flora, both among
phanerogams and cryptogams which have been named for Rev.
Moses A. Curtis, of North Carolina, a most acute botanist and a
gentleman whose character and ability reflect honor upon the
State in which he lived. Curtis was born in Berkshire county,
Massachusetts, in 1808, and at the age of 22 came to Wilmington, N. C., as a tutor in the family of Governor Dudley. He
began at once to devote himself to the study of the flora of that
region, which was especially rich and interesting. Close up to
the village reached the pine forests, abounding in strange plants
that charmed the eye and filled the portfolios of the enthusiastic
young botanist. His quick eye and assiduity may be judged by
the fact that in little more than two seasons during his brief hours
of leisure he made a collection of over a thousand species, or
rather more than half the number described in Elliott's Botany.
The result of these two years' investigation appeared in 1834, in
Curtis's " Enumeration of the Plants Growing Spontaneously
around Wilmington, N. C.," published in the Boston Journal of
Natural History. This first contribution to botany by Dr. Curtis
was more than a mere catalogue, and it attracted the favorable
notice of his teachers and correspondents. It was so thorough
that after the lapse of half a century only about 50 species have
been added to his list. Among some of the new plants which he
found near Wilmington was the curious and very local *Dionæa
muscipula*, or Venus fly-trap. Week after week he would visit the
savannas, and, lying at full length upon the ground, would watch
the peculiarities of this plant, and the description which he gave
of its habits in his first published work has been many times
quoted during the last 50 years, showing that he possessed the
gift of accurate and entertaining description to a marked degree.

In 1835 Curtis was ordained to the ministry of the Episcopal Church and immediately entered upon mission work in Western North Carolina. He spent a year in this work, and while thus engaged he took advantage of his journeying in the solitary woods to pursue his botanical researches. He traveled mostly in a sulky which was so arranged that his collecting portfolios could be placed under the cushion of the seat. As he came across specimens he would gather them, put them into the portfolios, and so by the end of his journey he had secured a goodly number of ready pressed plants for future study or for mounting in his herbarium. In 1849 he again visited the mountain region, and in 1841 it was said of him by Dr. Gray that " no living botanist was so well acquainted with the vegetation of the Southern Alleghany mountains, or has explored those of North Carolina more extensively."

Dr. Curtis' method as a student was that of a broad-minded scientist; just to name a flower and preserve it was to him but the beginning of his work. His earliest records show that he studied the relations of plant life to geologic and climatic surroundings. The study of botanical geography was begun with his career as a botanist and continued throughout, extending over 38 years. The account he gives us in his " Woody Plants " is to-day the best guide to the natural climatological divisions of the State which has ever been published. He also directed his attention to the numerous economic questions which met him in his intimate acquaintance with the treasures of the field and forest. It was this feature of his labors alone which brought him an audience in his adopted State. His " Woody Plants," published as a part of the State Geological and Natural History Survey in 1860, at once became a popular manual for the farmer and the woodsman, and for the amateur botanist a key to the more conspicuous trees and shrubs useful for their fruit or timber or as ornaments. The preface of this little work is an introduction to the geographical distribution of plants in the State and shows what a thorough acquaintance its author had with the broad subject.

Although Dr. Curtis is known as a man who was intimately acquainted with the flowering plants of the South, it is through his great knowledge of cryptogams, especially of fungi, that he became most widely known and justly famous. This very difficult branch of botany had few votaries in Curtis's time, and there was no text-book on the subject published in America. Provided, however, with the two well-known works of Schweinitz, Curtis addressed himself to what was for him a labor of love. He was painstaking and accurate in his methods, and the microscopic work necessary for the determination of species became with him a triumph of skill. Few were the botanists with whom he could compare specimens or exchange notes. He pursued this specialty without the stimulus now offered by special societies, and for the greater part of his career absolutely without an audience. It was intense love for the work which led him up to the highest station occupied by any American botanist. Dr. Curtis gave particular attention to the edible species of fungi. He communicated to the " Gardeners' Chronicle" for October 9th, 1869, an article on the " Edible Fungi of North Carolina," and left in manuscript a very complete illustrated work on " Esculent Fungi." Every one knows the palatable and wholesome character of the common mushroom or pink gill, but few are aware that there are other kinds growing in our fields and woods that are more finely flavored and just as wholesome. In his catalogue of the plants of North Carolina, Curtis indicates 111 species of edible fungi known to inhabit that State, and he remarks elsewhere that he has no doubt that there exist 40 or 50 more.

In a letter to Berkeley, of England, Curtis writes: " In October, 1866, while on the Cumberland Mountains in Tennessee, although with little leisure for examination during the two days spent there, I counted 18 species of edible fungi. Of the four or five species that I collected for the table, all who partook of them declared them most emphatically delicious. On my return home, while stopping a few hours at a station in Virginia, I gathered eight good species within a hundred yards of the depot. And so it seems to be throughout the country. Hill and plain, mountain and valley, fields and pastures, swarm with a profusion of good nutritious fungi which are allowed to decay where they spring up, because people do not know how, or are afraid, to use them."

In 1867 the State published as a part of the Geological and

Natural History Survey, a " Catalogue of the Indigenous and Naturalized Plants of North Carolina," by Dr. Curtis. This Catalogue included 4,800 species, and was the first attempt made by any botanist in this country to enumerate the cryptogams as well as the flowering plants, and its appearance was a matter of much scientific congratulation. This work was the result of 25 years of botanical study over a territory of 50,000 square miles.

The most important purely scientific papers of Dr. Curtis were his contributions to the mycology of North America, published in several volumes of Silliman's Journal.

A genus of plants, *Ravenelia*, belonging to the same family as the rust of wheat, perpetuates the name of Dr. Henry W. Ravenel, of South Carolina. At the time of his death, less than four years ago, there were among American botanists none more respected and beloved and few whose scientific work covered so many years of activity.

Born with a fondness for natural history, Ravenel pursued his botanical studies with great earnestness and enthusiasm during the whole of his long life. He made extensive collections of plants, but he was no ordinary collector, heaping up rough material to be exchanged for specimens to be counted rather than studied. He was a most accurate observer, and always noted the habits and peculiarities of the plants he gathered. There was not a group of plants, no matter how small, which escaped his observation. He not only studied critically the phanerogams of his State, but he collected and studied as far as it was possible in his time and in a region remote from large libraries, mosses, lichens, algæ and fungi. He was a zealous follower of Dr. Curtis, and became recognized as authority on the species of fungi in the Southern States, for no one possessed such an intimate acquaintance with them. He discovered many new flowering plants and a surprisingly large number of cryptogams. His interest in the latter group brought him into correspondence with the leading mycologists of Europe, and Ravenel became nearly as well known abroad as at home. He is best known to the botanists of to-day by his published sets of fungi, a dozen volumes or more, which are now rare and exceedingly valuable. His writings are not numerous, but they are characterized by thoroughness and indicate an active mind which went beyond mere descriptions and inquired into causes as well as results. His work was not solely for the scientific world; he was for many years botanist to the State Board of Agriculture and an editor of an agricultural paper. His popular articles on the grasses of the South are of special interest and value, for he devoted much attention to these plants and appreciated their agricultural importance. He is spoken of as having been a man whose life was full of kindness to all about him and an example of what a botanist should be.

Among the Southern botanists who have worked in this field and become identified with Southern plants are many who have attained distinction through their labors, but of whom I have not the time to speak. I must, however, mention the names of Lindheimer and Fendler, whose collections have become classical through the publication of Engelmann and Gray; of William F. Feay, of Georgia, whose devotedness to botany led to a number of interesting discoveries; of Prof. John L. Riddell, who for many years resided at Mobile, and whose name is indelibly associated with botany through the genus Riddellia; of S. B. Buckley, who discovered and published many new species of Southern plants, and for whom *Buckleya distichophylla*, a graceful Santalaceous shrub of the mountains of North Carolina was named; of Dr. C. W. Short, of Kentucky, justly famous in his time and whose services were recognized by Dr. Gray in that extremely rare plant found only near Roan Mt., *Shortia galacifolia;* of John Williamson, of Kentucky, the artist botanist whose "Fern Etchings" gained him a world-wide reputation; of Le Conte, of Georgia, who monographed our species of the genus *Paspalum;* and lastly, but by no means least, of Dr. Engelman, of St. Louis, Mo., whose botanical works placed him in the first rank of men of science and whose publications will always be essential to every working botanist.

Passing over these thus briefly, we come to those who are yet living and still active in the cause.

We cannot omit speaking of Judge M. Thomas Peters, of Moulton, Ala., although he is no longer actively engaged in botany. Judge Peters was one of the first to draw the attention of botanists to the ferns of his State and was the discoverer of that beautiful little species named in his honor, *Trichomanes Petersii.*

There are numerous private as well as public herbaria, both in this country and in Europe, that owe many of the treasures they possess to the kindness and generosity of Judge Peters. He brought to light that rare and curious sedge, *Carex Boottiana*, and he was one of the earliest students of the fungi of Alabama. His collections of this class of plants as well as his botanical library he donated to the Alabama State University.

The discoverer of the very rare *Neviusia Alabamensis*, the Rev. R. Nevius, should be named in 'passing, although he is now stationed on the Pacific coast. He communicated many interesting Southern plants to Dr. Gray, and in his new field of labor he is continuing his botanical studies, paying particular attention to mosses.

Louisiana is at this time the fortunate possessor of a most industrious and acute botanist in the person of Rev. A. B. Langlois, of St. Martinsville. Mr. Langlois was born in France in 1832, and he began his botanical studies in that country, for before coming to America in 1855, he had made an herbarium of some 1200 species. He spent nearly two years in Cincinnati completing his ecclesiastical education, and then located at Point-a-la-Hache, La., where he remained for thirty years. The locality being near the delta of the Mississippi was one of peculiar botanical interest, and Mr. Langlois succeeded in discovering many rare and some new species of plants. Langlois has carried on his botanical studies under circumstances which would have deterred many from undertaking them. He has been entirely cut off from botanical associates, and the climate of his region is so moist as to render the drying of specimens most difficult. Upon going to Point-a-la-Hache, he at once renewed his botanical work, but being entirely without books and wholly unacquainted with any American botanist, he sent his first collections, numbering some 300 species, to France to be named, but he never heard from them or received one word of encouragment. Evidently disheartened he dropped the study of plants for twenty years, a period which he now looks upon with deep regret. In 1878 he began again the study, first with only Wood's "Manual," and then with Chapman's "Southern Flora." Langlois thus relates his progress from this time, "By accident I learned that there

was a botanist, Dr. Puissant, at the Ecclesiastical Seminary of
Troy, N. Y., and I immediately wrote to him offering Southern
plants for Northern ones, and I received from the doctor about
500 species. Soon after I found out there was published here a
" Botanical Gazette," for which I immediately subscribed. From
this journal I learned many things unknown to me before;
through its advertisements I got plants from Eggert, of Missouri,
Pringle, of Vermont, and a check-list from Patterson, of Illinois.
Then I began to know and appreciate the advantages of having
correspondents. The ones who have been of greatest service to
me in phanerogams are Morong, of Massachusetts, Wibbe, of
New York, and later, J. Donnell Smith, of Baltimore. In grasses
I have been assisted by Dr. Vasey, of Washington, and in Cyperaceæ by Connant.

"In 1884, through the kindness of Mr. Lehnert, of Washington, I began the study of mosses, liverworts and lichens, and in
the latter part of 1885, at the suggestion of Mr. Scribner, I began
the study of fungi. I soon acquired a deep interest in these
plants, and have been greatly aided in their study by Prof. Ellis,
of New Jersey. The mycological flora of Louisiana being so rich
and at the same time so poorly known, I have for the past three
years given almost my entire attention to it. Every day I make
new discoveries, and I am yet far from having exhausted this intensely interesting part of the Louisiana flora." Mr. Langlois has
now an herbarium containing some 5000 species of North American plants, including 1214 species of Phanerogams and vascular
cryptogams of his State. So far as his State is concerned, this
work has been done single-handed. About a year ago, Langlois
published a catalogue of Louisiana plants which embraced the
fungi he had found, now numbering 1200 species. Langlois'
collections are widely distributed in the herbaria of this country
and in France, and his specimens are highly valued by all who
possess them.

I have been thus particular in speaking of Mr. Langlois, not
only to show the interest that may be acquired in the study of
botany, but also to show what may be accomplished under conditions most adverse. Mr. Langlois is now rector of St. Martin's
church, St. Martinsville, La.

For many years botanists of this country, when they have wished specimens of Southern plants or have desired to learn more about them, have turned by almost common consent to Dr. Chas. Mohr, a druggist of Mobile, Alabama, a German by birth. An assiduous explorer and collector, an acute observer of plants and a generous correspondent, freely communicating to others the results of his labors, Dr. Mohr has won the esteem and confidence of all American botanists. Nor is his reputation confined to this side of the Atlantic, for some of his botanical communications to the German scientific journals have been translated into all the leading languages of Europe. What he has accomplished for the science of botany has been done in the hours of recreation which he could command in a pretty hard struggle for an existence in his calling as pharmacist. During the first years of his residence in this country, from 1857 to 1865, he devoted the limited hours he could spare for rest or recreation to the study of mosses, and the specimens he collected of these plants were sent to the leading bryologists of this country and Europe. He greatly assisted Lesquereux and James by furnishing material for their work on the mosses of North America, and the many new species which he found were published in the BULLETIN OF THE TORREY BOTANICAL CLUB under the title of "Contributions to the Bryology of the United States."

In 1886 Mohr acquired the valuable herbarium of Dr. J. Riddell, of Louisiana, author of a "Catalogue of the Plants of Louisiana" and of a "Synopsis of the Plants of the West." This herbarium contained the extensive collections of Dr. Josiah Hale, of Alexandria, Louisiana, as well as those of Prof. W. M. Carpenter, and was to Mohr "a mine of wealth and information." At about the same time he had placed in his hands for determination the valuable collection of Dr. Denney, of Luggsville, La., from which he got a glimpse of the richness of the arboreal flora of Southern Alabama beyond the pine belt proper. The study of this collection inspired him with a special interest in woody plants and in forestry, and his work in this field of botany now stands second to none. The results of a season spent in the field of 1877 form the chapter on the "Forests of Alabama," published in 1878 in Birney's Handbook. This paper more systematically arranged, corrected, and new points of special interest to the subject added, was afterwards published separately.

For a time connected with the State Grange, Dr. Mohr directed his attention to the grasses and forage plants, with particular reference to those best adapted for cultivation in the coast region of the Gulf States. In this connection he first pointed out the value of Mexican clover and Lespedeza striata. His investigations on these plants were fully published in the reports of the U. S. Department of Agriculture. He also communicated frequent articles to the local papers on this and kindred subjects.

Many interesting waifs from foreign countries have been introduced on ship's ballast around Mobile, and the studies Mohr made of these plants led to the publication in the "Botanical Gazette" of a paper "on the plants introduced into the Gulf States." Further experiences on the same subject gained during the succeeding ten years furnished the theme for an article published last August in the "Pharmaceutical Revue" (German), on "Plant immigration in the Southern States through the aid of animals or accidently by man."

In 1880 Dr. Mohr was called upon by Prof. Sargent to undertake the investigation of the forests of the Gulf States, in regard to their timber resources and other products and the industries dependent on them. This work kept him engaged in the field during all of one season and part of another. The results of his observations were reported to Prof. Sargent and published in the ninth volume of the Tenth Census report. While in the field Mohr discovered several new species and rediscovered a number that had long been lost sight of. His observations also furnished him materials for an article on forest trees of the Gulf region published in Vol. I of the American Journal of Forestry, and another on "The Forests of the South and their Bearing upon the Interests of Agriculture," which was read at the meeting of the Southern Interstate Immigration Association at Nashville.

In 1878 Prof. E. A. Smith, of the State Geological Survey, placed in Dr. Mohr's hands the plants he had collected during his field work, and from that time dates Mohr's undertaking to get up an herbarium for the Survey to be placed in the museum of the University of Alabama. Uniting his own collections with those of Prof. Smith as a basis, the work has progressed so that now the herbarium contains something over 2200 species of flowering plants and ferns, nearly completely illustrating the entire State Flora.

In 1882 or 1883, Dr. Mohr was engaged by the Louisville R. R. Co. to collect and report upon the products of the forest and fields along the line of its road within the State. The collections made formed a most interesting and attractive feature in the Exposition held at Atlanta and at Louisville, and in 1884–85 they were again exhibited at New Orleans. An account of the material gathered was published in a pamphlet entitled "The Natural Resources of Alabama," one of the few papers of its kind which possesses real scientific merit and in no way can mislead the reader or prospective settler.

A paper on the "Resiniferous Pines of the South and the Manufacture of Naval Stores," published in the *Pharmaceutical Review*, attracted the attention of the present chief of the Forestry Division of the U. S. Department of Agriculture and led to the engagement of Mohr to prepare for the Department a series of exhaustive monographs on the Southern pines of economic importance, to form a part of a report to be devoted to the biology of North American timber trees.

Mohr is possessed with a true scientific spirit and great enthusiasm in his botanical work. By the amount he has accomplished it is very evident that he has well improved his hours of leisure and doubtless stolen much time from his hours of needed recreation. But in this day and generation one cannot stop to recreate, for if he does, some more zealous worker will win the prize he seeks. Success from true merit seems now to depend upon one's powers of endurance.

Mohr has the distinction of having gone out of the beaten track of systematic botanists and considered the plants he studied from an economic aspect. He has not only increased the sum of our knowledge, but he has added to our powers of direct usefulness. I would say to those who, in referring to botany, are ever asking the question "*cui bono?*" carefully read the writings of Dr. Mohr; they afford a most able answer.

A leguminous plant found only in Florida was named by Drs.

Torrey and Gray, *Chapmania*, in honor of him who is now the oldest of living botanists, Dr. A. W. Chapman, of Appalachicola, Florida.

On the 16th of May, almost sixty years ago, there landed at Savannah, Georgia, a young man from New England, who has since gained a world-wide reputation as a botanist and become more than all others identified with the plants of the South. The day following his arrival, in a stroll beyond the city limits he found that curious pitcher plant, *Sarracenia variolaris*, and with the aid of Eaton's "Manual" he determined its name. This was the starting point of a botanical career which culminated in 1860 in the production of Chapman's "Southern Flora."

In 1835 Dr. Chapman settled in Florida, where he has since resided, and during all these years his interest in or love for his chosen science has never for a moment flagged. He has always been isolated from association with botanists, if we except the brief companionship which he enjoyed with Mr. H. B. Croom, whose botanical services have been commemorated in the genus *Croomia;* but he has numbered among his correspondents all the prominent botanists in this country and many in Europe. Only seven of these were ever known to him personally, and he has outlived them all.

Dr. Chapman thus writes me, by request, concerning the origin of his principal publication—the "Southern Flora." He says:

" I believe the ' Flora ' owes its existence to a suggestion of mine to Dr. Curtis about the year 1856, that we needed for the South a work something like what Gray had made for the North, and that he (Curtis) was just the man to do it. But at that time his hands were full of toadstools and he was rusty among the Phanerogams, and so turned over the matter to me, while promising every assistance in his power if I would undertake the job. My time also was fully occupied with my professional duties during the day, but I concluded to try my hand at it, after office hours, by way of experiment. Well, I succeeded better than I had anticipated, and was encouraged to go on, and so night after night, from 9 to 12 or 2 o'clock, for nearly three years, until ' Finis ' was reached, found me at work. In the summer of 1859 I took my manuscript to Cambridge and consulted with Gray about

further proceedings, the result of which was that I concluded to
be my own publisher. I went to the University Press, selected
the neat typography in which the work is dressed and set the
printers to work—remaining with Gray, proof-reading and correcting until November, when I turned the concluding pages over to
Prof. Eaton and came home. The work was issued just before
the war began, and I heard nothing more of it for four years,
when Dr. Gray smuggled through the lines a budget of friendly
notices of the work which appeared during those years in the
periodicals of this country and Europe, and all at once I awoke to
bigness."

Since 1860 Dr. Chapman's Flora has been the standard botany
for the Southern States, ranking with Gray's "Manual" for the
States of the North. Early in 1883 Dr. Chapman published a
reprint of the work, to which he added a supplement in order to
include the many plants discovered within the range covered by
the Flora, particularly in Florida, since the publication of the first
edition. And now at the advanced age of 80 years, the doctor is
actively engaged in the preparation of another edition of his Flora
to meet the changes which have been made in recent years in
systems of arrangement and in nomenclature. It was only the
other day that I received a letter from him asking the loan of
some specimens from the Gattinger Collection in order that he
might settle for himself some determinations that had been
brought into question. I do not know how far this new edition
has progressed, but it is the hope of all botanists that this ven-
erable, most courteous and faithful worker may live to again write
the word finis and see the full fruition of his labor.

The rich and varied flora of our own State has been made
known to the world by Dr. A. Gattinger, of Nashville, through
his publication entitled "Tennessee Flora." Dr. Gattinger came
to this State some 42 years ago and was located for a time here
in the East, but he soon went to Nashville, where he has since
resided. During all these years he has been an industrious collector and close observer of Tennessee plants, making many journeys to inaccessible or out-of-the-way places to discover some
variety or increase his knowledge of plant distribution. Quite a
number of new species have been found by him and several of

these bear his name. His herbarium, of which the University of Tennessee is now the fortunate possessor, was one of the finest private collections in the country. It numbers between three and four thousand species, besides a large stock of duplicates which can be used in making further accessions to the herbarium proper. The Flora of Tennessee and the general Flora east of the Missis- sippi are well represented in it. The collection has been enriched through exchanges with many prominent botanists, both North and South.

Dr. Gattinger has directed special attention to the grasses of Tennessee, and he contributed the most important chapters in Killebrew's work on these plants.

His " Flora of Tennessee," published in 1887, is a work of excellent merit and at once stamps its author as a botanist of the first order. In speaking of his experience, in the preface to the " Flora," he says " I am yet in possession of specimens collected in 1849, when I first took up my residence in East Tennessee as a practicing physician. Placed as I was in those early days amid unfamiliar modes of life, with no access to intellectual resources, without information about the condition and advance of scientific affairs in this country, my botanical progress could for many years be no other than tedious and slow; but I kept up the pursuit, which since early school years had been to me a source of pleasure and consolation. His " Flora" is based upon botanical collections made exclusively by himself during 38 years' residence in the State. In it some space is given to a description of the physical characters of the State and the peculiar flora of each division, and the habitat and date of flowering of the species enumerated is given. To the students of botany within the State it is indispen- sable. It has not been my good fortune to meet Dr. Gattinger, but he is a most agreeable correspondent and his letters bespeak a man of high culture and refinement.

He has accomplished much by his industry and passionate love for his chosen work, and there are few names better known or held in greater esteem by the working botanists of the country than that of Dr. Gattinger, of Tennessee.

The list of Southern botanists is a long one, and contains many honorable and even brilliant names. I would that I were better able to speak of them as they deserve, but I have already put your patience to a test, and must now leave the subject with the feeling, which I know is shared by some of you, that "the half has not been told."

HISTORICAL SKETCH
OF THE BOTANY OF LOUISIANA

R. S. Cocks

VI.

HISTORICAL SKETCH OF THE BOTANY OF LOUISIANA.

R. S. COCKS,

(Communicated February 3, 1899).

The earliest printed notice of the Flora of Louisiana seems to have been in the year 1758, by the historian Le Page du Pratz. In the first chapter of the second volume of his history of Louisiana he states that he had spent nearly sixteen years in the study of the plants and animals of Louisiana, and that he had made such progress in the botany of the state that he had sent to France a collection of three hundred medicinal plants worthy of notice in the form of a report to the French government. Whether this collection is still in existence I have been unable to discover.

Le Page du Pratz does not claim to have made an exhaustive study of the botany of the state, but to have confined his attention to such plants as were supposed to have an economic value The first chapter of the second volume is on Grasses and Leguminous plants. He mentions several species of both, but rather of such as were found to flourish in the state, like corn and rice, than of species indigenous to the country. Chapter II is on fruit trees and enumerates a large number, both native and introduced.

Chapter III is on the forest trees, and contains notes on about fifty of our native trees and their uses. Chapter IV is a short description of a number of our native shrubs, and concludes with a short account of what he calls "Excrescences," under which term he includes mushrooms and lichens. There is also a short description of *Tillansdia Usneoides*, which he calls

"Barbe Espagnole," deriving its name from the strong re-
semblance which the Indians detected in it, to the beards of
the Spaniards.

Chapter V is on climbers and other plants—their virtues
and some flowers. Of the flowers he says that all kinds are
so numerous that the perplexed collector cannot make up his
mind which to take and which to leave behind when all are so
beautiful, and that they bloom from May to December. All
these chapters are copiously illustrated by quaint but rather
perplexing pictures. Altogether this book, though exceedingly interesting from an antiquarian point of view, would
not be of much practical use to the botanist of to-day.

The next published notice of any sort on the Botany of
the State was in 1802. A little book was published, entitled
"Travels in Louisiana," by a man named Davis. It is a translation from the French and contains a chapter "On the Trees
of the Colony." It does no more than give the names of a
few trees observed in the State.

In 1807 we have the first serious attempt to treat directly
of the Botany of the State in a work entitled "Florula Ludoviciana," by C. C. Robin. This gentleman published in Paris,
in 1807, in three volumes, his travels in Louisiana, West Florida and the West Indies, performed between 1802 and 1806.
At the end of the third volume he inserted a "Flora of Louisiana" classed according to the method of Jussieu. He describes
about four hundred species of plants, collected principally
near New Orleans, of which one hundred and ninety-six were
at that time new and forty-six were designated as useful.
He also suggests several plants as worthy of cultivation in
our gardens. His plants were made up as follows: Phanerogams 389, Fungi 7, Ferns 4.

In 1817 this book was translated into English, and enlarged and improved by Rafinesque. In his introduction he
says: "Whoever shall have the opportunity of comparing my
labors with the fragments of Robin will soon perceive the
wide difference between our works. The numberless defects
of his flora were of such a nature as effectually to preclude the
possibility of its being practically employed by a botanist. I
have endeavored to collect names and describe the objects
which he had merely pointed out and generally drowned in
useless and superfluous details."

He goes on to state that in order to render this "Florula"
less incomplete he has added in a supplement an enumera-

tion of all the plants already mentioned by authors as natives of the State of Louisiana and omitted by Robin.

These authors are principally Bartram, Michaux, Pursh and Muhlenberg. These additions bring up the number of plants known in Louisiana to 596. This little book may be fairly taken as summing up all that was known of the botany of Louisiana at that date.

In the same year, 1817, was published by William Darby, a work entitled "A Geographic Description of the State of Louisiana." This book contains a chapter on the "Vegetation Peculiar to Louisiana," but does not add very much to our knowledge of the subject.

In the year 1825 there was published in French a description of the United States by D. B. Warden. The article on Louisiana contains a list of the trees and shrubs of the state, 101 in number; also the plants used in medicine [4] and plants "good for the nourishment of beasts" [7]. A considerable number of these are not contained in the Florula Ludoviciana. The plants which he mentions as being medicinal are *Alisma odorata, Panax quinquefolium, Solidago odora* and *Prenanthes serpentaria.* The forage plants consist of three species of clover, one fern and two grasses, and also the wild strawberry.

From the year 1825 we have to pass over twenty-seven years to the year 1852, in which year was published in the New Orleans Medical and Surgical Journal of 1852 and 1853, a catalogue of the plants of Louisiana, by Prof. Riddell of New Orleans, Dr. Josiah Hale of Alexandria, and Prof. Carpenter of New Orleans. Prof. Riddell, who was professor of Chemistry in the University of Louisiana, is also famous as the inventor of the binocular microscope. These three gentlemen did great work as pioneers in the Botanical exploration of the state. Their catalogue contains the names of over 1500 species of flowering plants and thirty species of vascular cryptograms. Grasses and sedges are not included in this catalogue, but mention is made of an additional list of these groups by Dr. Hale, which I have not been able to find or hear anything of. In the introduction to the catalogue Riddell states that, "this list embodies the results of a great many years of observation by Dr. Josiah Hale, Prof. Carpenter and the author, and has been abridged from a work in manuscript, contributed by the author in 1851 to the Smithsonian Institution." The work alluded to is entitled "Plants of Louisiana."

It comprises the technical and vulgar names of the flowering plants and ferns well ascertained as growing in the limits of Louisiana, with special localities, times of flowering and full descriptions of new species. The Cyperaceæ and Gramineæ, contributed by Dr. Hale, are not included in the present abridgement.

Again for a long period there seems to have been a lull in botanical work in Louisiana, until in 1863 appeared a book, entitled "Resources of the Southern States, Fields and Forests," by Francis Porcher, which contains a few notes on some Louisiana species but nothing of much importance. I might also have mentioned an article in the "American Journal of Science" of 1848 on Southern plants by Dickeson and Brown.

The next contributor to the Botany of Louisiana was Prof. A. Featherman of the State University at Baton Rouge. His collections were not very extensive compared with those of Riddell, and were confined principally to the vicinity of Baton Rouge. The results of his labors were published in three reports of the Botany of Louisiana in 1869, 1870 and 1871.

In 1871 the second edition of Alphonso Wood's Class Book of Botany was published. This book is of great assistance to students of the Botany of the State; for he tells us in the introduction that while he was working on the book, Dr. Hale sent him a set of specimens representing the Flora of Louisiana, and that descriptions of these are incorporated in his book.

Up to this date, 1871, Botanists in Louisiana had confined themselves almost entirely to flowering plants and ferns. The year 1887 introduces us to the pioneer worker in the cryptogamic botany of the State. In this year the Rev. A. B. Langlois, now of St. Martinville, La., an honored member of our society, published "A Provisional Catalogue of the Phanerogams and Cryptogams of Louisiana." This catalogue he tells us in the introduction, was the result of eight years collecting, principally in the neighborhood of Plaquemine parish.

It stands as a monument to the untiring zeal, industry and ability of its author. The list contains the names of 1194 phanerogamic plants, 22 vascular cryptogams, 103 species of mosses, 30 species of hepaticæ, 663 species of fungi. Many species of all these were new to science. While his catalogue of flowering plants is not so extensive as that of Riddell's, it contains a good many species not mentioned in the former.

Contemporary with Father Langlois, was another very able and diligent botanist, Dr. Joor, for several years the curator of the Tulane museum. Though the greater part of his botanical work was done in Texas, he spent several years investigating the flora of Louisiana.

Through the kindness of Professor Trelease and Prof. J. B. S. Norton of the Missouri Botanical Gardens, I have been supplied with a catalogue of his collections. This collection which is now in the possession of the Missouri Botanical Gardens is not very large, as ill health and his professional duties at Tulane prevented his doing as much work in collecting as he would otherwise have done, but I find in it several plants not mentioned by any of the others. There is a very interesting account of his life and work in the Botanical Gazette of October 1898.

In 1892 in the Geological Survey of Louisiana, published at Baton Rouge, were two articles on the Botany of the state, one on North Louisiana, the other on South Louisiana.

The second one is by Prof. Dodson of the State University and Mr. Mathews, and gives a list of the principal plants of economic value in the Florida parishes. It enumerates about three hundred plants.

The one on North Louisiana is by Mr. Thomas Wayland, and gives a list of trees and shrubs found and determined in North Louisiana. It enumerates about 200 species; also twenty medicinal plants and fifty-seven grasses.

During the last two years Prof. John K. Small of Columbia University has been publishing in the Torrey Bulletin, a series of "Studies in the Botany of the Southern States." In these studies are descriptions of a number of species from Louisiana. The author has very liberally presented the Society with a full series of these.

This I believe about ends the list of all the botanical workers in our State; of all those at all events who have published anything. If there are any whose names I have omitted I would be very glad to hear of them.

With this brief sketch I submit to the Society "A preliminary catalogue of the flowering plants and ferns of the state*." The grasses and sedges are not included in this but will be added at a later date. This catalogue may be taken as fairly well summing up the labors of all the earlier workers, and of

*Owing to the great number of additional species catalogued since the reading of this article, I have thought it advisable, with the concurrence of the Executive Committee, to postpone the publication of the check-list.—R. S. C.

Riddell, Hale, Carpenter, Dr. Joor, and to a certain extent, the work of Father Langlois up to 1887. There are about 1800 plants enumerated. Of these I have collected myself something over three-fourths during the last two or three years, and have been enabled to add about sixty species to the lists of the others, which are recorded here for the first time as occurring in the state. Most of the remaining fourth can be found in the Tulane Herbarium. Those I have not seen are inserted upon the authority of Dr. Joor or taken from the 1889 catalogue of Rev. A. B. Langlois.

This catalogue is to be regarded as merely a preliminary one, to be followed at a later date by a more complete annotated one, in the preparation of which we hope to have the assistance of the Rev. A. B. Langlois who, from his long experience in the state and his wide knowledge of the subject, is far better prepared than any one else in the state for such a work.

Bibliography of Botany of Louisiana.

1758. History of Louisiana—Le Page du Pratz, Vol. II.

1802. Travels in Louisiana—Davis.

1817. Florula Ludoviciana—C. C. Robin and Rafinesque.

1817. Geographical description of the State of Louisiana—William Darby.

1828. Description of the United States—D. B. Warden.

1852. Catalogus Flora Ludovicianæ—auctore J. L. Riddell [in New Orleans Medical and Surgical Journal].

1863. Resources of the Southern States, "Fields and Forests."—Francis Porcher.

1869 and 1870. Featherman's reports—Geological Survey of Louisiana.

1871. Text book of Botany—Alphonso Woods.

1888. Catalogue Provisoire de Plants. Phanérogames et Cryptogames—Rev. A. B. Langlois, St. Martinville, La.

June, 1890. New Species of Louisiana Fungi—Ellis and Langlois. [Journal of Mycology].

1892. Geological Survey of Louisiana—Two articles on Botany of the State.

1898. Studies in the Botany of the Southern United States in the Torrey Bulletin—John K. Small.

A BIOGRAPHICAL HISTORY
OF BOTANY AT ST. LOUIS, MISSOURI

Perley Spaulding

[Parts] I-IV

A BIOGRAPHICAL HISTORY OF BOTANY AT ST. LOUIS, MISSOURI[1]

By Dr. PERLEY SPAULDING

LABORATORY OF FOREST PATHOLOGY, BUREAU OF PLANT INDUSTRY,
U. S. DEPARTMENT OF AGRICULTURE

THE history of botany in St. Louis extends back nearly to the beginning of her political history. The city was founded in 1764, and while it is not as old as most of the other large cities of this country it seems to have been one of the earliest settlements made in the great northwestern region, comprising what was once known as Upper Louisiana. Boston, New York and Philadelphia were already large cities for that time and were centers of botanical activity. In 1795 when Michaux visited the Illinois Territory, Cahokia, Kaskaskia and St. Louis were the principal places west of Vincennes and as late as 1800 St. Louis had a population of less than 1,000. At about this time the fur traders changed their headquarters from Cahokia and Kaskaskia to St. Louis, causing a corresponding increase in population and commercial influence of the latter town.

The Jesuit missionaries were the first white persons to visit the Mississippi Valley and the adjoining country; they undoubtedly explored the Missouri Territory, but probably not so extensively as they did the Illinois Territory. They were versed to some extent in the art of medicine and knew the plants which were generally used for medicinal purposes. They learned the uses of plants new to themselves from their Indian wards, and in this way they must have obtained a considerable knowledge of the plants of the Missouri country. How much farther they may have carried their botanical studies is unknown to the writer. During the period between the founding of St. Louis and the first visit of Michaux to Cahokia there were undoubtedly persons who studied the botany of the St. Louis district. Whether they formed any collections of the plants is not now known and there seems to be no records of any such study.

For all practical purposes André Michaux may be said to have been the first botanist to work in the vicinity of St. Louis.

Botany has passed through a number of distinct periods at St. Louis, as in other places; it can not be said to have had a "pharmaceutical" period, as that stage was nearly past in the general history of the science when the city was founded. The medical properties of

[1] Published by permission of the Secretary of Agriculture.

the plants of the eastern states and of Europe were already well known at the close of the eighteenth century. Of course many new western plants were discovered, the medical properties of which had to be determined; but this was not the main object in making a study of them. We find three distinct periods of botanical work which include the one hundred and thirteen years that have elapsed since Michaux's visit. These may be designated as follows: First, exploration by botanists on transient visits of a few days' to a few months' time; second, collecting by persons who lived in or near St. Louis for a number of years; third, modern botany as contrasted with the purely systematic work of early days. These three periods overlap one another, but can still be distinguished without difficulty. The first includes most of the work done previously to 1850; the second began with the work of Engelmann and his numerous contemporary collector friends, who relied upon him for assistance in naming their collections; it may even be said to extend until the present time, as considerable work is still being done upon the local flora of the district; the third period may be said to date from the founding of the Shaw School of Botany, and the assumption of control of the Missouri Botanical Garden by the board of trustees.

André Michaux, the great French botanist, who explored so extensively the territory of the thirteen original colonies as far west as the Mississippi River, is the first botanical worker concerning whom published records have yet been found as having worked in the vicinity of St. Louis. He is known to have visited Kaskaskia and Cahokia, and the evidence seems to indicate that he must have visited the west shore of the Mississippi, since a few species are listed in his " Flora " as coming from the Missouri River.

André Michaux[2] was born at Satory, near Versailles, France, in 1746. He was destined by his father for the superintendence of a farm of the royal estate, and early became interested in agriculture. Upon the death of his young wife, at the birth of their son, François André, he devoted himself to scientific studies, especially botany. He studied botany under Bernard de Jussieu, and sought in foreign lands for strange plants. In 1779–81 he traveled in England, the Auvergne, the Pyrenees and Spain. In 1782–5 he was in Persia in a political capacity, but really to explore a country at that time almost unknown to scientific men; he intended to return to Persia, but was requested in 1785, by the French government, to introduce into France such North

[2] Hooker, W. J., *Amer. Jour. Sci. and Arts*, 1st series, 9: 266–269, 1825.
Gray, Asa, Ditto, 1 ser., 42: 2–9, 1842.
Coulter, J. L., *Bot. Gaz.*, 8: 181–183, 1883.
Rusby, H. H., *Bull. Torrey Bot. Club*, 11: 88–90, 1884.
Sargent, C. S., " Scientific Papers of Asa Gray," 2: 23–31, 1889.
Thwaites, R. G., " Early Western Travels," 3: 11–19, 27–104, 1904.

American trees as might be of economic importance. In the autumn of 1785 he embarked for New York, accompanied by his young son; here he spent a year and a half collecting plants and starting a botanical garden in Bergen County, New Jersey; he found, however, that the southern climate was more suitable for many of his plants, and he accordingly removed to Charleston, South Carolina, in 1787, where he established another garden, about ten miles from the city. During this year he explored the mountains of the Carolinas; the next he journeyed through the swamps of Florida, and the next he visited the Bahamas, and again searched the mountains for plants of economic importance—especially ginseng. In 1792 he collected around New York and in New Jersey; thence he went up the Hudson to Albany and along Lake Champlain, reaching Montreal June 30, 1792. From Montreal he went to Quebec, and thence by way of the Saguenay to Hudson's Bay. He then returned to Philadelphia, where he proposed to the American Philosophical Society an exploration of the great western territory, by way of the Missouri River. A subscription was begun for the purpose, and Thomas Jefferson drafted detailed instructions for the journey. Michaux, indeed, is stated to have started west and to have proceeded as far as Kentucky when he was overtaken by an order from the French government to relinquish the journey for a political mission. This mission seems to have had for its object the control of Louisiana by the French, through the aid of the trans-Allegheny Americans. In carrying out this plan Michaux made a journey in 1793 to Kentucky by way of the Ohio River, and returned over the "Wilderness" road, and through the valley of Virginia. Early in 1794 he made another extensive tour in the southern states and the North Carolina mountains. In 1795-6 he made a much longer journey, going from Charleston to Tennessee, thence through Kentucky to Vincennes, Indiana, where he stayed from August 13 to 23. From here he went to Kaskaskia, and from there he visited Cahokia and the vicinity. Upon looking over his "Flora Boreali Americana" we find several species of plants mentioned therein as coming from the Missouri River. It seems quite probable then that he must have visited some locality near this river during this trip, as this is the only visit to this section of which we find any mention in his journal. He mentions St. Louis as being in a prosperous condition, but makes no further allusion to it. Except for the evidence of these few species as given in his "Flora," we should not know that he had gone west of the Mississippi River, and this, of course, is somewhat uncertain, as it is very possible that some person at Cahokia, who may have been on the Missouri River, had out of curiosity picked up some strange plants and happened to bring them to Cahokia at the time Michaux was there. He made a short visit here and then went to Fort Massac, near the mouth of the Tennessee River, and from there proceeded up the Cumberland River by boat, as far as Clarksville; he then visited Nashville, Knoxville, Louisville, and Morganton, finally arriving at Charleston again in April, 1796. During all of this time he collected eagerly, and more or less extensively. His journals, however, give no indications of the species or the number of them found at Cahokia. He seems to have found a considerable number at Kaskaskia, at which place he spent most of his time while in Illinois. In his " Flora " we find mentioned about 100 species as occurring in the Illinois territory; this, however, at that time included all of the territory north of the Ohio which was visited by Michaux. This seems to have been his last extensive trip in America; and in August, 1796, he embarked for Amsterdam and was shipwrecked on the coast of Holland. He is said to have been nearly drowned himself, and a large part of his collections were lost. He remained in France for several years, studying his collections and preparing the manuscript for his " Flora." In 1800 he joined an expedition to Australia, but, becoming disgusted with the management, he landed on the Island of Mauritius, but from there he soon went to Madagascar; here he established a botanical garden and began collecting extensively; but he soon fell victim to the unhealthy climate, and died on November 13, 1802.

Michaux probably traveled more extensively in North America than any other early botanist. He was the author of numberless new species and many new genera of American plants. Unfortunately, the genus, *Michauxia,* which commemorates his name, is one discovered by himself in Persia; so that his name is not thus associated with North American botany, which was so greatly advanced by his studies and explorations.

Immediately following the exposition held at St. Louis in commemoration of the purchase of Louisiana from France, there was held another exposition upon the Pacific coast to celebrate the centennial of the arrival of the Lewis and Clark expedition at the mouth of the Columbia River. This expedition was the first to penetrate overland to the Pacific coast and the results of its successful termination were of immense importance to the entire northwestern country. The journals of the expedition contain many references to plants seen, and especially to those which were peculiar or interesting, or which were used by the Indians.

In the previously mentioned attempt at the exploration of the northwest country, Michaux was to accompany the party. In the expedition which finally did make the journey there was no person who could be called a botanist. Although Captain Lewis was a very keen and observant man, he could not overcome his lack of botanical training, and the results in this regard were hardly what they would have been had Michaux been with the expedition. The journey up the Missouri River was made in boats manned with oars and, owing to the rapid current of the river, progress was slow, thus affording opportunity for a considerable amount of collecting to be done. During the ascent of the river quite an extensive collection of plants was made, but this had to be left behind when the Rocky Mountains were crossed, and was consequently lost. During the much more hurried return of the expedition another collection was made, but it was much smaller than the first, and comparatively few species seem to have been collected about St. Louis. While this expedition did but little for St. Louis botany directly, it turned the public attention to this section, and finally led to careful botanical exploration by a number of capable botanists a few years later.

Captain Meriwether Lewis[3] was born near the town of Charlottesville, Virginia, on August 18, 1774. His family was one of the most

Fig. 1. Captain Meriwether Lewis; from *Analectic Magazine and Naval Chronicle*, Vol. 7, 1816.

distinguished of that state. Several of his uncles were very prominent in their time, one of them having married a sister of George Washing-

[3] Jefferson, Thomas, "Biography of Capt. Lewis in Analectic Magazine and Naval Chronicle," 7: 329-333, 1816.

Allen, Paul, "History of the Expedition under the Command of Captains Lewis and Clark," etc., 1814, reprint by New Amsterdam Book Company.

ton. Meriwether lost his father early in life, and one of his uncles acted as his guardian. At the age of thirteen he was sent to the Latin school, where he remained until he was eighteen, when he returned home to help run the farm. At the age of twenty he entered as a volunteer a body of militia which was called out by General Washington to quell troubles in the western states, and from the militia he entered the regular service as a lieutenant. When twenty-three years old he was promoted to a captaincy and made paymaster of his regiment. He was personally well known to Thomas Jefferson, and when the latter proposed that two persons should be sent up the Missouri River, across the Rockies and down the Columbia to the Pacific Ocean, he eagerly offered to go. A few years later Jefferson, remembering the eagerness of Captain Lewis to make the trip, made him leader of the expedition, which successfully carried out the plans, and is now known as the Lewis and Clark Expedition. Captain Clark was made the leader in the absence of Lewis. The expedition started in 1803 and returned in 1806. Congress gave both leaders grants of land, and Lewis was made governor of the territory of Louisiana, while Clark was made a general of militia and agent for Indian affairs. Upon assuming his duties as governor, Lewis found many factions and parties, but his even-handed justice to all soon established respect for himself, and eventually removed animosities. While on a trip to Washington he suffered a temporary attack of insanity, and committed suicide on October 11, 1809.

Pursh has named a genus of the Portulacaceæ, *Lewisia,* in his honor.

During the early part of the nineteenth century it was much the fashion for botanists to collect living plants and cultivate them in gardens, these gardens sometimes being quite extensive. Sometimes they were but temporary resting places for the plants until they could be sent to European countries as novelties to be introduced there because of some desirable quality. André Michaux had such gardens into which he gathered his plants, and when opportunity offered sent them to France. Many of our early botanists had their own gardens in which they cultivated all of the different plants they could find, and thus became acquainted with every detail concerning them. The Bartram and Marshall gardens near Philadelphia were good examples of these early collections of living plants.

Among many persons sent from Europe to this country for the purpose of collecting new and rare plants was one John Bradbury,[4] who was commissioned to act as the agent of the Liverpool Botanical So-

[4] Bradbury, John, "Travels in the Interior of America in the Years 1809, 1810 and 1811," 1-346, 1819, 2d edition.

Short, C. W., *Transylvania Jour. of Med.,* etc., 34: 12-13, 1836.

Britten, Jas., and Boulger, G. S., "Biographical Index of British and Irish Botanists," 21, 1893.

ciety. Comparatively little seems to be known about Bradbury. He was a Scotchman who had lived for a long time in England, when he received his commission from the Liverpool Botanical Society in 1809. Upon arriving in this country, Bradbury spent several days at the house of Thomas Jefferson, so that the latter became acquainted with him and his abilities. Jefferson spoke highly of him as a naturalist, and Short, a later writer, mentions him as "an English gentleman of very respectable attainments as a naturalist." In the light of our present knowledge he seems to have fully deserved such an estimation, as he discovered a considerable number of new species as well as a new genus of plants during his travels in the Missouri country. Indeed, several of our more characteristic species bear his name, and in later years he was honored by Torrey and Gray, who named a new genus *Bradburia,* in commemoration of his services in exploring our western flora.

Mr. Bradbury at first intended to make New Orleans his center of operations, but following the advice of Jefferson he changed that intention and came to St. Louis instead. He descended the Ohio River by boat, making such observations and collections as he could at the various stopping places, arriving at St. Louis on the last day of the year 1809. The entire season of 1810 was spent about St. Louis, making short excursions of not more than eighty or one hundred miles distance in all directions, and he accumulated a considerable collection of plants which were sent to Liverpool the succeeding autumn. No definite data can now be obtained as to the number of species contained in these collections, as Bradbury never published a complete list of them, although he did give a list of the rare and more interesting plants in his journal, which was published after his return to England.

Early in the spring of 1811 Bradbury, accompanied by a young and zealous botanist named Thomas Nuttall, joined a fur-trading expedition, and with them ascended the Missouri River as far as the Mandan villages, not far from the site of the present town of Bismarck, North Dakota. Upon reaching this point the expedition divided and part of it, including Bradbury, returned to St. Louis. The others went on still farther, and Nuttall remained with them until their return to St. Louis some months later. This voyage was made in a steamer, and progress was necessarily slow while going up the river, so that our naturalists had ample time and opportunity for collecting. A collection even larger than that which had been made around St. Louis is said to have been accumulated.

Before Bradbury had finished his preparations for departure to England, the war of 1812 broke out, and he remained for several years in this country until the close of hostilities. He finally reached Liverpool in 1815, and found that during his long absence his plants had been inspected by Pursh, who was at that time in England preparing the manuscript for his " Flora Americanæ Septentrionalis." Pursh published the most interesting of these plants in an appendix to his work, and this seems to have discouraged Bradbury from publishing as extensively upon them as he probably would have otherwise done. In 1817 Bradbury published his journal of travels on the Missouri in the years 1809–10–11, and in an appendix to this gave a list of the rare and most interesting plants of his collections. He did not, however, issue a complete list, and so far as now known no such list has ever been published. The second edition of his travels was issued in 1819, and in the editor's preface it is stated that Mr. Bradbury had already returned to St. Louis and taken up his residence there. Baldwin, who passed through St. Louis in 1819 with the Long expedition, mentions meeting Mr. Bradbury there at that time. His name is given in the St. Louis city directory for 1821,[5] but no definite information regarding him after this date has yet been found.

During the early part of the nineteenth century it was the policy of the national government to send expeditions of a military character to explore the unknown sections of the western country. Shortly after Bradbury made his tour of the Missouri, an expedition was fitted out and placed under the command of Major S. H. Long. This was intended to make more complete and detailed exploration of the Missouri and its main tributaries, and to make more accurate scientific observations of the country passed through. The necessity of having competent scientific men accompany the expedition was recognized, and several such men were appointed for the purpose.

The botanist of the expedition was Dr. William Baldwin.[6] He was a son of a minister of the Friends in Pennsylvania, being born in Newlin, Chester County, in 1779. He studied medicine at the University of Pennsylvania and took his degree in 1807. Meanwhile he had become interested in the study of botany, and upon settling in Wilmington, Delaware, to commerce practising his profession, he collected extensively in the vicinity. Pulmonary weakness forced him to remove to Georgia in 1811, where he served as surgeon to a gunboat flotilla during the war of 1812. He kept up his collecting and study of the plants of this new region, and because of his ability as a botanist he received an appointment as surgeon to the U. S. frigate *Congress,* during a cruise to various South American ports. Baldwin made extensive collections and notes wherever opportunity offered, and he returned with a

[5] Learned by the aid of the St. Louis Historical Society.
[6] Thwaites, R. G., "Early Western Travels," Vol. 14.
Darlington, William, "Reliquiæ Baldwinianæ," 1–346, 1843.
Redfield, John, *Bot. Gaz.,* 8: 233–237, 1883.
Harshberger, J. W., "Botanists of Philadelphia," 119–125, 1899.

very considerable amount of valuable material. About this time the Long expedition was being organized, and Baldwin was recommended to act as botanist for the party. His health was delicate and the appointment was accepted in the hopes that it would be improved by the journey.

Baldwin joined the other members of the scientific staff at Pittsburg and embarked upon the steamer which was to take the expedition to Council Bluffs. This being the early days of steamboating, the one used by the expedition gave more than ordinary trouble and caused

FIG. 2. DR. WILLIAM BALDWIN; from Darlington's " Reliquiæ Baldwinianæ."

vexatious delays. According to the letters of Baldwin it also leaked continually, and this made the interior damp and totally unsuited for such a prolonged voyage. Baldwin's health constantly grew worse, and even while descending the Ohio River the party halted to allow him to recover from an attack of illness, and he was forced to depend upon the others to bring specimens to him on the boat, as he had not sufficient strength to walk any considerable distance. St. Louis was finally reached on June 9 and a stop of several days was made. The voyage was resumed on the twenty-first, and on July 13 they reached Franklin, then the uppermost town of any importance on the Missouri. Here Baldwin was left behind at the house of Dr. Lowry, where he remained until his death on August 31. During his stay in Franklin Baldwin botanized as much as his limited strength would permit, and entries were made in his diary as late as August 8, the date of the last entry. A list of plants found around Franklin by him during this time attests the earnestness with which he pursued his beloved science. The journals of the expedition show that he collected about one hundred species in the vicinity of St. Louis and on the Missouri to Franklin.

His companions all unite in praise of his devotion to science and his persistence under such extremely trying circumstances. Notwithstanding his extensive travels and his earnest study of the botany of several different sections of this country and of South America, he published but little. Two short articles, presented for publication just before starting with the expedition, are all that are known to have been published by him. He left numerous manuscripts and notes which have aided Torrey and Gray in their work on the flora of America. His herbarium was extensive and very valuable, and has contributed much to the works of Pursh and Nuttall. Baldwin also contributed to Muhlenberg's catalogue, and he maintained an active correspondence with many of the foremost botanists of his day. Nuttall has honored him by naming a genus of the Compositæ *Baldwiniana,* and has thus connected him in a most permanent manner with that science to which he so earnestly devoted himself.

The Long expedition proceeded and on September 17 went into winter quarters near Council Bluffs. Major Long meanwhile went east, and on his return brought with him Dr. Edwin James, who had been appointed to take the place of Dr. Baldwin.

Edwin James[7] was born in Weybridge, Vermont, on August 27, 1797. Edwin was the youngest son of Deacon Daniel James, who was a native of Rhode Island, and had moved to Vermont at the beginning of the Revolution. In youth he was very industrious and applied himself to his studies with perseverance. His education was obtained at the district school, and later he attended Middlebury College, where he graduated in 1816. Subsequently he studied medicine with his elder brother in Albany, New York, for three years. During this time he became interested in botany and the natural sciences, which were then being taught by Professor Amos Eaton. Upon the recommendations of Captain Lewis Le Conte and Dr. John Torrey he was appointed to the place left vacant by the death of Dr. Baldwin. The trip with Major

[7] Thwaites, R. G., "Early Western Travels," Vol. 15.
Parry, C. C., *Amer. Jour. Sci. and Arts*, 2d series, 33: 428–430, 1862.
Sargent, C. S., "Silva of North America," 2: 96, 1891.

Long was a hurried one, although it was made overland from St. Louis to Council Bluffs and but few plants were collected near St. Louis. James remained with the expedition until its close. His efficient labors are proved by the subsequent publications founded upon his observations and collections. The present Pikes Peak was first named James's Peak, by Major Long, but for some unexplained reason the earlier name has not remained in use.

The next two years after the return of the expedition were spent in compiling his results, which were published in 1825, and were of much historical and scientific value. During the next six or seven years he served as a surgeon in the regular army at extreme frontier posts, and here he studied the Indian languages and translated the New Testament into the Ojibwe tongue. He also published a biography of John Tanner, a man who was captured by the Indians while a child, and was brought up by them. When the medical department of the army was reorganized he resigned and returned to Albany, where he was associate editor of a temperance periodical. Upon leaving this he went west and settled near Burlington, Iowa, where he spent the last days of his life in agricultural pursuits. On October 25, 1861, he was run over by a wagon and injured so seriously that he died three days later.

The genus *Jamesia*, of the Saxifrage family, was named in his honor by Torrey and Gray.

The results of the exploring expeditions seem to have directed attention to the Missouri country, so that a number of men of ability came to that section and made botanical explorations of greater or less extent. Before the Long expedition had finished its work an amateur botanist, Dr. Lewis C. Beck, was collecting about St. Louis.

Dr. Lewis Caleb Beck[8] was born in Schenectady, New York, on October 4, 1798. In 1817 he graduated at Union College; he then studied medicine and began to practise at Schenectady in 1818. He moved to St. Louis in 1820 and lived here until 1822. During this time he collected quite extensively and later published a list of his collections. His introductory note is self-explanatory and is as follows:

During my residence in Missouri, in the years 1820, 1821 and 1822, a portion of my time was occupied in the investigation of the vegetable productions of that and the adjoining state. Upon my return I was so fortunate as to receive, uninjured, the collections which I had made.

Until the present season (1826), however, I have not had leisure to examine them with the necessary attention, and to revise my notes upon the recent plants. This work I have now commenced, and submit to you the first part,

[8] Appleton's "Cyclopedia of American Biography," 1: 213, 1887.
Anonymous, *Amer. Jour. Sci. and Arts*, 2d series, 16: 149, 1853.
March, Dr. Alden, Gross's "Amer. Med. Biography," 679–696, 1861.
Beck, L. C., *Amer. Jour. Sci. and Arts*, 10: 257–264, 1826; 11: 167–182, 1827; 14: 112–121, 1828.

for publication in your valuable journal. Those species which are presented as new are minutely described, and in all cases where the western specimens of known plants differ from the eastern, this difference is stated. By this means we shall become acquainted with, at least, some of the peculiarities in the vegetation of that interesting section of the United States. Concerning the more common plants, the habitats and times of flowering only are mentioned. The catalogue, it is hoped will contribute somewhat to increase our stock of knowledge, and will be particularly interesting to geographical botanists, and to future writers upon the botany of the United States.

This annotated list, which was continued in three volumes of *Silliman's Journal*, mentions about two hundred species of plants, and is the earliest extensive list known to the writer. Many of Beck's plants are cited in Riddell's " Synopsis of the Flora of the Western States," published in 1835, but apparently only a portion of them are so mentioned.

In 1822 Beck moved back to Albany and remained there the rest of his life. He held positions as professor of botany and other sciences at a number of institutions up to the time of his death; Rensselaer Polytechnic Institute, Rutgers College and Albany Medical College being those with which he was most prominently connected. Dr. Beck was well known in botanical circles, being the author of a manual of the botany of the northern and middle states, of which two editions were issued. He also published a number of botanical papers. He was a well-known writer on chemical and medical subjects besides; and published a manual of chemistry which passed through four editions. He seems to have been a conservative writer, as his bibliography contains but twenty-three titles. Dr. Beck died at Albany on April 20, 1853.

After Beck closed his work in the vicinity of St. Louis there seems to have been a period of nearly ten years when there was no botanical work done. In 1831, however, there began a period of activity which has continued more or less regularly up to the present time. The first botanist to start this activity was Thomas Drummond.

(To be continued)

A BIOGRAPHICAL HISTORY OF BOTANY AT ST. LOUIS, MISSOURI. II.

By Dr. PERLEY SPAULDING

LABORATORY OF FOREST PATHOLOGY, BUREAU OF PLANT INDUSTRY,
U. S. DEPARTMENT OF AGRICULTURE

THOMAS DRUMMOND[9] was born about 1780. He is known to have been a native of Scotland, but the exact place of his birth is unknown, as is also his early training and education. He was a brother of James Drummond, the Australian botanical explorer, and is known to have succeeded George Don in the nursery at Forfar. In 1825–6–7 he accompanied the Second Overland Arctic Expedition, led by Sir John Franklin, as assistant to Dr. Richardson, who was the naturalist of the expedition. In Canada Drummond explored very extensively, even into the Rocky Mountains and on the Mackenzie River where the main part of the expedition did most of its work. Upon the completion of the journey he returned to England, and from 1828 to 1829 he was curator of the Belfast Botanical Garden. Soon after his return to England he published a work upon the American mosses, which was chiefly the result of his collections made in Canada. He again sailed for New York under the patronage of Drs. Hooker and Graham, for the purpose of exploring the southern and western United States. Beginning his tour at New York City in the spring of 1831, he went to Philadelphia, visited Bartram's garden, thence to Baltimore, Washington, and to Wheeling on foot. At the last-named place he embarked for St. Louis, descending the Ohio River and coming up the Mississippi by boat. It was his original intention to join some fur-trading expedition to the far western country, but he arrived in St. Louis too late for this. He accordingly remained in St. Louis and collected in the vicinity until the next winter. He lost considerable time by sickness, but in January he sent a collection of several hundred species of phanerogams and a considerable collection of mosses and hepatics to Hooker at Kew. Hooker

[9] Date of birth and photograph supplied by Mr. J. R. Drummond, grandson of Thomas.

Hooker, Wm. J., *Companion to the Jour. of Bot.*, 1: 21–26, 39–49, 95–101, 170–177, 1835; 2: 60–64, 1836. *Journal of Botany*, 1: 50–60, 183–202, 1834. *Botanical Miscellany*, 1: 178, 1849.

Lasègue, A., "Musée Bot. de M. Benj. Delessert," 196–198, 1845.

Sargent, C. S., "Silva of North America," 2: 25, 1891.

seems to have prepared his collections for distribution, and we find him publishing a list of about two hundred and fifty species which were collected around St. Louis by Drummond.

During the next spring and summer Drummond collected in the vicinity of New Orleans, and here he obtained even more plants than he did at St. Louis. He next went to Texas, which he was one of the first to explore botanically. Here he gathered a rich harvest, in spite of a season of the most unfavorable weather. He then returned to New Orleans and went to Appalachicola in 1835 for the purpose of exploring the Florida peninsula. He soon left western Florida with the intention of reaching Key West by way of Havana, Cuba. Hooker learned that Drummond was taken sick while at Havana and died very suddenly in March, 1835.

Harvey dedicated the genus *Drummondita* to the two brothers.

[9a] By an unfortunate error, this portrait of Thomas Drummond was in the last issue of the MONTHLY printed as a portrait of William Baldwin, and the portrait of William Baldwin was printed as the portrait of Meriwether Lewis.

FIG. 4.   PRINCE MAXIMILIAN : from his " Reise nach Brasilien."

Very appropriately *Drummondia,* a genus of American mosses, was named in Thomas's memory by his patron, Sir William Jackson Hooker. Numerous species of our phanerogams are also named after this most industrious and successful collector.

Even persons of royal lineage were numbered among the many naturalists who came to America for the purpose of exploring unknown sections for new plants and animals, and to make scientific observations. While Prince Alexander von Humboldt attained eminence for his travels and scientific worth, he was not the only royal person who did so, although we generally hear no other mentioned. Alexander Philip Maximilian, Prince of Wied neu-Wied, came to the New World on two different occasions. On the first tour he visited Brazil, and on the second he visited the United States and especially the northwestern or Missouri country.

Prince Maximilian[10] was born on September 23, 1782, in Wied

[10] Thwaites, R. G., " Early Western Travels," Vols. 22, 23 and 24.

Maximilian, Prince, " Reise nach Brasilien," 1820-1.

Sargent, C. S., " Silva of North America," 9: 138, 1896.

neu-Wied, a small principality of Rhenish Prussia. He was from boy-hood of a studious inclination, and early became interested in the natural sciences. In spite of this he was in the Prussian army at the battle of Jena, and was among those captured by the enemy. He returned to his studies at the end of this war, but was among the victorious army which entered Paris in 1813. In this service he earned the iron cross of Chalons and a major-generalship. During all of this time he had been planning a scientific expedition to Brazil in order to satisfy a keen de-sire to add to the world's knowledge, imparted to him by the celebrated Professor Johann Friederich Blumenbach, of whom he was a favorite pupil. Early in 1815 he started for Brazil. He was joined in South America by two other German scholars, and the trio spent two years studying the flora, fauna and native races of this country. His result-ing publications gave him a high rank among the scientists of the period, and his " Reise nach Brasilien in den Jahren 1815 bis 1817 " was soon translated into the French, English and Dutch languages.

In 1832 Prince Maximilian started on a second enterprise—a trip to the trans-Mississippi region. He arrived in Boston on the fourth of July. He brought with him a very capable artist, for the express pur-pose of obtaining portraits of famous Indians. He made more or less brief visits to Boston, New York and Philadelphia, and then went to Bethlehem, Pennsylvania, and thence through the coal region, reaching Pittsburg in the autumn. The journey was then continued overland to Wheeling, where they embarked for the voyage down the Ohio River. They turned aside for the purpose of visiting New Harmony, Indiana, where then was located the best library of American and natural history west of the Atlantic seaboard. Here the winter was spent studying and preparing for the journey on the Missouri River. On March 16, 1833, the journey was resumed and they arrived in St. Louis before the fur-trading expeditions had left on their annual trip to the northwest. Following the advice of several St. Louis men, the journey was made by boat up the Missouri River, instead of by land, as was at first planned. On April 10 the journey was commenced, and by the twenty-second they had reached Fort Leavenworth. The expedition was continued to Fort McKenzie, on a branch of the Yellowstone River, among the Blackfeet Indians, where they remained for two months. The return trip was be-gun on September 14, and the succeeding winter was spent at Fort Clark, near the present town of Bismarck, North Dakota. The next spring Prince Maximilian returned to St Louis and journeyed eastward by way of the Ohio canal and Lake Erie to New York, where he em-barked for the Old World on July 16, 1834. Upon returning from the upper Missouri country the collections which had been made were left behind to be sent down the river in another steamer which was soon to follow the one carrying the party. A fire broke out on this steamer and many of the collections were destroyed because they were not deemed of as much value as other things which were on board.

After his return to his native city Prince Maximilian worked over his collections and other material with the aid of a number of experts, and published several papers upon his results. In 1843 he published his "Systematic View of Plants Collected on a Tour on the Missouri River." His collections are preserved in the museum of his native city, where he died in 1867.

Martius honored him by naming a genus of Brazilian and West Indian palms, *Maximiliana,* thus very appropriately connecting him with the botany of that country, of which he was one of the pioneer explorers.

Hardly had Prince Maximilian started for home before another explorer was at work on the Missouri. This person was none other than Thomas Nuttall, the greatest botanist of this country in his time. As has been already mentioned, he had visited this section in company with John Bradbury in 1811.

Thomas Nuttall[11] was born in the town of Settle, England, in the year 1786. His parents were in very moderate circumstances, and the boy was early apprenticed to a printer. After several years he had a disagreement with his employer and went to London seeking for work. Here he came very near total destitution. When about twenty-two years of age he emigrated to America, landing in Philadelphia. During his youth he so improved his spare moments that he acquired an intimate knowledge of the Latin and Greek languages, and he seems to have studied other branches, as he was described at the time of his landing as "a well-informed young man, knowing the history of his country, and somewhat familiar with some branches of natural history, and even with Latin and Greek." Nuttall knew nothing of botany at this time, but very soon after he became interested in the "amiable science," and also began an acquaintance with Dr. Benjamin Smith Barton. His studies of plants naturally led him into making short excursions which soon lengthened as his interest deepened, until he had visited the lower part of the Delaware peninsula and the coast region of Virginia and North Carolina.

At about this time Nuttall became acquainted with John Bradbury, and he eagerly proposed to accompany him on his trip up the Missouri River. Accordingly, Nuttall joined Bradbury at St. Louis, and early

[11] Short, C. W., *Transylvania Jour. of Med.,* etc., 34: 14-16, 1836.

Meehan, Thos., *Gardeners' Monthly,* 2: 21-23, 1863.

Durand, Elias, *Proc. Amer. Phil. Soc.,* 7: 297-315, 1860.

Sargent, C. S., "Silva of North America," 2: 34, 1891.

Britten, Jas., and Boulger, G. S., "Biographical Index of British and Irish Botanists," 129, 1893.

Anonymous, POP. SCI. MONTHLY, 46: 689-696, 1895.

Harshberger, J. W., "Botanists of Philadelphia," 151-159, 1899.

FIG. 5. THOMAS NUTTALL; from THE POPULAR SCIENCE MONTHLY, Vol. 46, 1895.

in the spring of 1811 the two made the journey up the river to the
Mandan villages, as previously described in this paper. Both naturalists were more than once in extreme danger, but Nuttall brought back
with him many treasures of seeds, plants and other objects of interest.
The next eight years he remained at Philadelphia, studying in the
winter the collections which he made during his summer excursions to
various parts of the country east of the Mississippi, from the Great
Lakes to Florida. At about this time he was preparing the manuscript

FIG. 6. ENTRANCE OF THE LINNÆAN GREENHOUSE AT THE MISSOURI BOTANICAL
GARDEN, ST. LOUIS, MISSOURI ; showing the three busts of
Linnæus, Gray and Nuttall.

for his " Genera of the North American Plants," which did not appear
until 1818. Nuttall himself set most of the type for this work. In
1818 he started from Philadelphia for Arkansas, reaching Fort Belle-
point on April 24, 1819. He made this his center of operations, exploring in various directions and making large collections. He was
taken sick with fever and on recovering made one more excursion and
then set out for home, reaching New Orleans February 18, 1820. At
this time he had made a journey of over five thousand miles through a
country still in the undisputed possession of the Indians, and almost
wholly unexplored by scientific men. Immediately upon his return to
Philadelphia in 1820 he began to study his collections and to write his

Fig. 7. Bust of Thomas Nuttall over the entrance of the Linnæan Greenhouse at the Missouri Botanical Garden, St. Louis, Mo.

"Journal of Travels into the Arkansas Territory During the Year 1819," which was published the following year.

At the end of 1822 he was called to Harvard College as curator of the botanical garden, there not being enough money to support a professorship. He soon became dissatisfied with this and took up the study of ornithology, producing a two-volume manual of this science. About the beginning of 1833 Nuttall went to Philadelphia with the collection of plants made by Captain Wyeth during an overland journey to the Pacific Ocean. A second journey was to be made and Nuttall resigned his position and spent the interval before the departure of the expedition studying the Wyeth collection and his own Arkansas plants.

Mr. Nuttall went in company with John K. Townsend, the two

FIG. 8.   GRANITE OBELISK IN THE MISSOURI BOTANICAL GARDEN NEAR THE MUSEUM
in honor of Thomas Nuttall.

being sent by the American Philosophical Society.   The two arrived
at St. Louis on March 24, 1834, on the steamboat *Boston*, from Pittsburg.   They started from St. Louis, going on foot to the point of
rendezvous at Boonville, Mo., where they joined the Wyeth party.
The brief period while they went on foot from St. Louis to Boonville
is the one which concerns us at present.   Unfortunately the season
was so early that Nuttall found but few plants in bloom.

The expedition ascended the Missouri River to the headwaters of
the Columbia, and then followed that to its mouth.   When winter
came on with our travelers on the Pacific coast they took passage for
the Sandwich Islands, where they arrived January 5, 1835.   Here

Nuttall remained for two months collecting plants and sea shells upon the different islands. He then separated from his companion and sailed for California. He spent most of the spring and summer upon the Pacific coast and then returned to the Sandwich Islands, where he embarked upon the same vessel that Dana was serving his " Two Years Before the Mast," to come home by way of Cape Horn. He arrived at Philadelphia in October, 1835, and settled down to study his treasures. For several years he worked thus and published two important memoirs. At Christmas, 1841, Nuttall went back to England, where he resided the last seventeen years of his life. This was not from choice, but because of the conditions under which an estate was left to him by his uncle, requiring him to live in England nine months of the year. He used his ample grounds for growing rare plants. Just previously to leaving the United States. he wrote a supplement to Michaux's " Sylva." In the preface his wanderings were outlined. He returned to America but once, when he took the last three months of 1847 and the first three months of 1848. At this time he studied the plants brought by Gamble from the Rocky Mountains and Upper California, and published a paper upon them. His death occurred on September 10, 1859, resulting from overstraining himself in opening a box of plants.

Torrey and Gray dedicated a genus of the Rosaceæ *Nuttallia*, to this prince of scientists.

Henry Shaw has honored him by placing a small obelisk of granite near the north end of the museum building in the Missouri Botanical Garden, with the following inscriptions: on the north side, " In Honour of American Science," and on the south side, " To the Memory of Thomas Nuttall, born in England 1786 and died September, 1859. Honour to him the zealous and successful naturalist, the father of western American botany, the worthy compeer of Barton, Michaux, Hooker, Torrey, Gray and Engelmann." He also placed over the entrance of the main greenhouse in the Garden three busts: that of Linnæus in the middle, and those of Nuttall and Gray on either side.

Although Nuttall explored the Missouri country on two different occasions and worked in Arkansas, he seems never to have published any considerable list of plants found by himself near St. Louis.

(*To be continued*)

A BIOGRAPHICAL HISTORY OF BOTANY AT ST. LOUIS, MISSOURI. III.

By Dr. PERLEY SPAULDING

LABORATORY OF FOREST PATHOLOGY, BUREAU OF PLANT INDUSTRY,
U. S. DEPARTMENT OF AGRICULTURE

EXPLORATION in the Missouri country was commenced in 1835 by Karl Andreas Geyer, a collector who became well known for his botanical explorations in the northwestern section of the United States. His explorations extended over a number of years and ranged from Illinois westward to the Pacific. He traveled especially in the territory included between the Mississippi and the Missouri River as far north as North Dakota.

Karl Andreas Geyer[12] was born in Dresden, Germany, on November 30, 1809. His father was a market gardener of very moderate circumstances. The boy was naturally bright and studied Latin under the tutelage of a kind-hearted man who helped him with his lessons, which were studied while he was selling his father's produce in the streets of the city. In 1826 he entered the garden at Zabelitz as an apprentice. In 1830 he removed to Dresden and engaged as assistant in the botanic garden there. In this place he had numerous friends, among whom was Dr. H. G. Reichenbach, whose lectures upon botany he attended with great regularity. He seems to have been a very likable and attractive person, drawing the attention of those with whom he came in contact. In February, 1834, he left Dresden for America. Here he collected plants during the summer months and worked at odd jobs in the winter, thus maintaining himself for several years. In one case he entered a newspaper office as compositor, but a few months later he was writing the leading articles for the same paper that he had helped set in type.

Geyer's first great journey in this country was in 1835, when he visited and explored the plains of the Missouri with a single companion. In 1836 and the succeeding years he went with Nicollet surveying the country between the Missouri and the Mississippi River. In 1840 he collected around St. Louis and in Illinois, making very considerable collections during this season. While in St. Louis he became acquainted with Dr. George Engelmann and this friendship seems to have lasted as long as Geyer was in this country. Engelmann seems to have worked over his collections, as we find him publishing upon them

[12] Anonymous, *Chronik des Gartenwesens*, 3: 185–187, 1853.
Reichenbach, H. G., *Kew Garden Miscellany*, 7: 181–183, 1855.

in 1841. He also came into possession of some of Geyer's collections, as it is definitely stated that they had been deposited in the Engelmann herbarium.

In 1841 Geyer went with Fremont to the Des Moines River in Iowa territory, where he found a number of new plants. In 1842 he explored the upper Illinois territory and formed the herbarium which was first offered for sale. In 1843 he began the journey from Missouri to the Pacific coast, lasting through the years 1843 and 1844. He explored the northwestern country very extensively and penetrated to hitherto inaccessible places by accompanying missionary trains on their visits to the different Indian tribes. He finally reached Fort Vancouver, and from there sailed on November 13, 1844, for England, going by way of the Sandwich Islands and Cape Horn. He arrived in England May 25, 1845, and spent some months at Kew, working over his collections and sorting out small lots of plants to sell. A large part of his profits from such sales was used in defraying expenses caused by a sickness brought on by his previous hardships. In September, 1845, he again returned to his home in Saxony, after an absence of eleven years. At first he entered the employment of head-gardener Lehman in Dresden, and later in the Royal Botanical Garden. His wanderings had shown him the value of a home, and on August 24, 1846, he married Miss Emma Schulze. Besides his duties for the garden he taught students the English language, his pupils coming from every class in Meissen. Geyer also took a prominent part in the local society for the advancement of science. During the last three years of his life he was editor of *Chronik des Gartenwesens und Feuilleton der Isis,* a periodical published at Meissen on the first and fifteenth of the month, from January 1, 1851, to December 15, 1853. Geyer's death occurred just before the end of the third volume, and it was discontinued with the third volume. While in no wise neglecting his duties at the garden, he came in written communication with the prominent botanists of the time and rounded out his collections. Heart disease troubled him considerably in his latter days and finally caused his death on November 21, 1853.

In 1835 a physician, George Engelmann by name, settled in St. Louis and soon built up a lucrative practise. During his spare moments he worked upon botanical problems, and before long he had established a reputation among botanists such that at his death he was ranked among the foremost of botanical workers.

Dr. George Engelmann[13] was born at Frankfort-on-the-Main, Feb-

---

[13] Gray, Asa, *Proc. Amer. Acad. Arts and Sci.,* 19: 516–522, 1884.
Sander, Enno, *Trans. St. Louis Acad. Sci.,* 4: 1–18 (Supplement).
Anonymous, Pop. Sci. MONTHLY, 29: 260–265, 1886.

FIG. 9.   DR. GEO. ENGELMANN; by courtesy of the Director of the
Missouri Botanical Garden.

ruary 2, 1809. He was the eldest of thirteen children. Aided by a
scholarship he went to the University of Heidelberg in the year 1827,
where he had as fellow-students and companions Karl Schimper and
Alexander Braun. Political embarrassments caused him to go in the
autumn of 1828 to Berlin University for two years, and finally to
Würzburg, where he took his degree of Doctor of Medicine in the summer of 1831. His inaugural dissertation, " De Antholysi Prodromus,"
published at Frankfort in 1832, testifies to his truly scientific mind.

Trelease, Wm., and Gray, Asa, " Botanical Works of Engelmann," 1–548,
1887.
Sargent, C. S., " Silva of North America," 8: 84, 1895.
White, C. A., " Biogr. Mems. Nat. Acad. Sci., 4: 3–21, 1896.

It is a morphological study founded chiefly upon monstrosities, and it had the honor of receiving the notice and approval of Goethe, who offered to place in Engelmann's hands his notes and sketches, which intention was frustrated by his death before it had been carried out. This first paper has been very favorably commented upon, and compared with much more extended and pretentious works of a similar nature.

The spring and summer of 1832 were passed at Paris in medical and scientific studies with Braun and Agassiz as companions. He then became the willing agent of his uncles, who had resolved to make some land investments in the Mississippi Valley, and he sailed from Bremen for Baltimore in September. He joined some of his relatives

FIG. 10. RESIDENCE OF DR. GEO. ENGELMANN IN ST. LOUIS, MISSOURI: by permission of the Director of the Missouri Botanical Garden.

who had previously settled in Illinois near St. Louis, and made lonely journeys on horseback through southern Illinois, Missouri and Arkansas. He finally established himself in St. Louis as a doctor of medicine late in the autumn of 1835. At this time St. Louis was a frontier town of eight or ten thousand inhabitants. Beginning in poverty, he soon built up a large practise and so established himself in his profession that he was able to go back to Germany for some months. While there he married his cousin, Miss Dora Hartmann, in June, 1840.

Again in 1856 he left his practise for a two years' absence, devoting the first summer to botanical investigations at Cambridge, and then visiting his native land in company with his wife and son. In 1868 the family again visited Europe for a year, the son remaining to study at Berlin. The mother died in January, 1879, and Englemann's own health failed alarmingly. A journey to Germany was taken in 1883 and the voyage was so beneficial that he was able to resume his botanical work. Serious symptoms soon caused him to return and the ocean voyage again proved very restorative and he resumed his labors with increased vigor. Increasing infirmities, however, gradually reduced his working powers until his death, which took place on February 4, 1884.

Upon first coming to this section of the country Dr. Engelmann traveled on horseback through southern Illinois and in Missouri and Arkansas; and during the latter part of his life he explored the mountains of North Carolina and Tennessee, the Lake Superior region and the Rocky Mountains and contiguous plains in Colorado and adjacent territories, thus being able to study in place, and with the acuteness of judgment which characterized his work, the Cacti, Coniferæ, and other groups of plants which he had investigated for years. In 1880 he made a long journey through the Pacific states, where he saw for the first time growing naturally many plants which he had described and studied over thirty years before.

Dr. Engelmann's papers are voluminous even for a man who could devote all of his time to botany; but it must be remembered that he had a large practise as a physician, which took most of his time, and that botany was taken up only in spare moments. When this is taken into account, together with the fact that he was also interested in other sciences (especially meteorology), their extent is nothing short of marvelous. The memorial volume of his papers published by Henry Shaw contains eighty-seven different papers of varying length. These have been grouped in this volume under the following headings or general topics: Cuscutineæ, Cacteæ, Juncus, Yucca and Agave, Coniferæ, Oaks, Vitis, Euphorbiaceæ, Isoetes, Miscellaneous, Lists and Collected Descriptions of Plants, and General Notes. It was the custom of Dr. Engelmann to take any scrap of paper and make notes upon it which might occur to him, together with sketches showing characters of the plant in hand. All such notes were at his death collected and mounted in a set of large books which are now in the possession of the Missouri Botanical Garden. These notes were so numerous that they made a library in themselves, filling sixty of these books.

His method of working was to take a single group of plants and work it out systematically so far as was in his power. His treatment of the genus *Cuscuta* in his first monograph of that group increased the number of species from one to fourteen without going west of the

Mississippi Valley. Seventeen years later, after an investigation of the whole genus in the principal herbaria of this country and of Europe, he published a systematic arrangement of all the Cuscutæ, giving seventy-seven species, besides a number of varieties.

Dr. Engelmann's authority upon the *Cactaceæ* was of the very highest. He established the arrangement of these plants upon floral and carpological characters. This work was carried on through a series of papers beginning with his sketch of the botany of Dr. A. Wislizenus's expedition from Missouri to northern Mexico, and continued in his account of the giant cactus of the Gila, in his synopsis of the Cactaceæ of the United States, and in his two memoirs upon the southern and western species contributed to the Pacific Railroad Reports and to Emory's " Report on the Mexican Boundary Survey." He had made preparations for a revision of at least the North American Cactaceæ, but upon his death much knowledge of this difficult group was lost.

His papers on the American oaks and the Coniferæ are of the highest interest, and are some of the best specimens of his botanical work; and the same is also true of his study of the vines. Nearly all that we know of this genus scientifically is directly due to Dr. Engelmann's investigations.

His work is characterized by a minuteness and carefulness of observation, coupled with a nicety of discrimination which made him a master in systematic work, his treatment

Fig. 11. Nicholas Riehl; from a photograph kindly loaned by his son, Mr. E. A. Riehl.

of the yuccas and agaves, the genera *Juncus, suphorbia, Sagittaria, Isoetes*, the *Loranthaceæ, Sparganum* and *Gentiana* giving him an eminence among fellows botanists to which few attain. His name was upon the rolls of many societies devoted to the investigation of nature, and he was the recognized authority upon those departments of his favorite science which had most interested him. His name has been given to a monotypical genus of plants, *Engelmannia,* by Torrey and Gray. Numerous species also bear his name.

Shortly after Dr. Engelmann settled in St. Louis, Nicholas Riehl.

a native of France, came to his city and settled on a piece of land on the Gravois Road in South St. Louis, and began to collect botanical specimens.

Nicholas Riehl[14] was born in Colmar, province of Alsace, France (now Germany), about 1808. His father's business was that of manufacturing cloth; not liking it, Nicholas sold it after the death of his father, and divided the estate. He took his share and traveled over much of Europe and America, coming as far west as St. Louis. Taking a liking to this part of the country, he returned to his old home and married. The two returned to St. Louis in the spring of 1836, and settled on a piece of ground on the Gravois Road in Carondelet, just outside the St. Louis city limits, and established a nursery. This is believed to have been the first nursery in St. Louis county, if not in the state of Missouri. The nursery business he carried on with success and profit until the time of his death in September, 1852. Riehl evidently collected botanical specimens some years before he came to this country, as specimens in his herbarium bear dates as far back as 1830, which were collected in the vicinity of Colmar. He also collected considerably in the vicinity of St. Louis in 1838. He had printed labels made for the collections made in this year, and they number not far from two hundred. Besides the specimens bearing the printed labels, there are many with incomplete labels which undoubtedly were collected here also. His entire collection was sold to Mr. Henry Shaw, who was at that time just starting to develop his botanical garden. The larger part of them were collected in Europe or were exchanged with European collectors. Mr. Riehl was a friend and admirer of Dr. George Engelmann, and was much interested in the work which he was doing. The Riehl nursery furnished Mr. Shaw the first trees which he planted in his newly started botanical garden.

In the forties Theodore C. Hilgard was collecting the native plants of the vicinity of St. Louis.

Theodore Charles Hilgard[15] was born at Zweibrücken, Rhenish Bavaria, on February 28, 1828. His father, Theodore Erasmus Hilgard, was a lawyer, who in 1836 resigned from the Supreme Court of the province and emigrated with his family to America, settling on a farm near Belleville, Ill., which at that time was the home of many other educated Germans who for political reasons had preceded him. Theodore was the sixth of a family of eight. The schools being poor and few in number, Theodore with the other younger children received his primary education from his elder sisters and elder brother Julius,

[14] Information and photograph supplied by Mr. E. A. Riehl, of Alton, Illinois, son of Nicholas.

[15] This sketch is adapted with very slight changes from a manuscript kindly furnished by Professor Eugene W. Hilgard, brother of Theodore.

while all received their higher training, especially in the languages, from their father. The boys aided in the farming operations and Theodore early manifested a marked interest in the natural sciences, and especially in botany; in which, however, his father could not help him. He soon found an enthusiastic helper in his younger brother Eugene, and together they made extensive collections of the native plants and insects of the vicinity. Dr. George Engelmann, a second cousin, greatly assisted the boys in their botanical studies.

Early in 1847 Theodore went to Europe and entered the University of Heidelberg as a student of medicine. Henle, Chelius and Hasse then made Heidelberg the most notable center for medical study outside of Vienna, while Bischoff represented botany. Hilgard at once began to make what subsequently became a very complete collection of the flora of central Europe. The revolutionary agitation of 1848 somewhat disturbed the regularity of the course of study, but no actual interruption occurred until, in the spring of 1849, active revolutionary movements took place in Baden itself. Theodore then (with his brother Eugene, who had meantime joined him) went to Zürich, and there passed three semesters, studying especially microscopy under Naegeli, and physiology under Ludwig, besides attending the natural history lectures of Oken. During this time the brothers made extended excursions on foot through Switzerland and collected the Alpine flora. In 1851 Theodore went to Vienna to study, where were then such medical celebrities as

FIG. 12. DR. THEODORE C. HILGARD; by courtesy of Dr. Eugene Hilgard.

as Rokitansky, Oppolzer, Bednar and Hebra. After nearly two semesters, during which he gave much time to botanical study in the great Endlicher collection, he was obliged to go to Malaga to bring back his widowed sister. While there he made an extensive collection of Mediterranean plants which greatly interested him. On his return he went to Würzburg, where he graduated in June, 1852, *summa cum laude*, as doctor in medicine, surgery and obstetrics. He then went to Berlin to study ophthalmology with Graefe, as well as surgery. In the summer of 1853 he returned to America, taking a position as ship physician on an emigrant vessel, on which he experienced an epidemic of cholera.

Soon after his arrival he went on a visit to the west to see whether he had best practise his profession there. On the way he sustained a severe shock to his spine in a steamer accident. It took him several weeks to recover somewhat, but he never fully recovered. He was disappointed in the outlook and returned to the east, where he took up practise in Philadelphia. There he became a friend of Elias Durand, a druggist and botanist, who in the latter capacity was requested to elaborate the botanical collections made by Heermann while with the Williamson Pacific Railroad Expedition. Durand proposed to Hilgard that they should collaborate in this work, and the latter being by nature an expert draughtsman, he not only described, but drew the illustrations of a large number of the " Plantæ Heermanianæ " accompanying the final report of the expedition. The strain of this work seemed to develop the spinal injury into a serious inflammation, from which he was prostrated for months. After recovery which was, however, never complete, he resolved to begin practise in St. Louis, and removed there in 1855.

He continued to practise in St. Louis from that time until 1870, much handicapped by the spinal weakness which obliged him to refuse much lucrative practice. His spare time was chiefly devoted to botanical studies, now more especially to the cryptogams, whose development he studied under the microscope, in the use of which he became very expert. In these studies he found that the then current classification and nomenclature of these organisms was seriously at fault, many merely developmental forms being classed as separate species, genera and even orders. He also worked zealously in devising a system of arrangement of the phanerogams which would express their mutual cross relations, the best graphic presentation of which on a flat surface he found in the pentagrammatic form. Comparative anatomy and the homotaxy of organs and structural parts also formed a favorite subject of investigation. Most of his work on these subjects was published in the *Proceedings* of the St. Louis Academy of Sciences, of which he was a charter member; also in the *Proceedings* of the American Association for the Advancement of Science and in the *St. Louis Medical Reporter.* He also helped in the organization of the " Humboldt Institute " library which for some time had a very useful cultural influence. In 1865 he married Miss Georgina Koch, daughter of Mr. A. Koch, of *Zeuglodon* fame. No children came of this union.

As the state of his health precluded his acting as an army surgeon, he remained at St. Louis during the war in hospital and private practise. After the war medical practise seemed to become more and more incompatible with his strength, and he gave it up and joined his brother Eugene at the University of Mississippi, where at that time a lectureship of botany was contemplated. But it failed of realization, and he accepted a position in the U. S. Coast Survey as observer in the magnetic survey then being made on the basis of the " Bache Fund." In this he continued until 1873, when he found it necessary to settle to a quiet life. The last year of his life was passed at New York City, where in March, 1875, he died of an abscess of the lungs.

The Hilgard collection of plants, embracing about 12,000 species, was taken by his brother Eugene to the University of California, where it was destroyed by fire in 1897.

Shumard in his presidential address before the St. Louis Academy of Sciences in 1869 spoke as follows concerning a collection of plants given by Hilgard to the Academy:

> Our botanical collection embraces an extensive series of lichens and mosses amounting to several hundred species, chiefly from western states and territories. These were collected by Dr. T. C. Hilgard, of this city, and by him presented and arranged in our museum.[16]

In the fire which destroyed part of the academy museum a few years later, this collection was also destroyed.

[16] Shumard, *Trans. St. Louis Acad. Sci.*, 3: XII., 1869.

(*To be continued*)

A BIOGRAPHICAL HISTORY OF BOTANY AT ST. LOUIS, MISSOURI. IV

By Dr. PERLEY SPAULDING

LABORATORY OF FOREST PATHOLOGY, BUREAU OF PLANT INDUSTRY,

U. S. DEPARTMENT OF AGRICULTURE

ONE of the best known of the botanical collectors of this country who worked shortly after the middle of the last century was August Fendler. He, like numerous others, came to America from Germany in the late thirties. From 1864 to 1871 he lived at Allenton, Missouri, about thirty miles from St. Louis. While living at Allenton Fendler arranged the first botanical specimens in the herbarium which was just being started by Henry Shaw for his Botanical Garden. These numbered about 60,000 and consisted of the herbaria of Bernhardi and Riehl, the latter containing a considerable number of local species. Because of his extensive and excellent collections, he became known to botanists and botanical institutions. While he was widely known by reputation, he seems not to have been well known personally, because of his excessive diffidence.

August Fendler[17] was born August 10, 1813, in the town of Gumbinnen, in eastern Prussia. When he was six months old his father died, and two years later his mother married again. His parents had but scanty means and his school training for a number of years could scarcely be called schooling. When about twelve years old he was sent to the Gymnasium, and was here for about four years, when his parents were obliged to take him from school because of financial troubles. He was apprenticed to the town clerk's office, and here began to think of traveling in foreign countries.

At the end of his apprenticeship he had an offer to accompany a prominent physician as his clerk in a journey of inspection along the Russian frontier of Prussia where the cholera was beginning to be feared. Fendler was soon in the midst of the cholera and remained for some time, returning home when the disease had abated. He now learned the trade of tanning and currying during the next two years. In the fall of 1834 Fendler was admitted to the Royal Gewerbeschule, but the strain upon his already frail health caused him to abandon it after finishing the first year with credit.

[17] Canby, W. M., *Bot. Gaz.*, 9: 111–112, 1884; 10: 285–290, 301 304, 319–322, 1885.

Gray, Asa, *Amer. Jour. Sci. and Arts*, 3d series, 29: 169–171, 1885.

Sargent, C. S., " Silva of North America," 12: 123–124, 1898.

In the fall of 1835 he started with a knapsack upon his back from Berlin as a traveling artisan, passed through parts of Silesia, Saxony, to Frankfort, down the Rhine, and finally coming to Bremen. Early in the spring of 1836 he embarked for Baltimore, Maryland, arriving with but two dollars in his pocket. In Philadelphia he worked in a tannery

FIG. 13. MR. AUGUST FENDLER, at about the time he lived at Allenton, Mo.

for a time, then went to New York and worked at the lamp manufacturing business. The financial panic of 1837 caused this business to be closed in the spring of 1838.

Having made up his mind to go to St. Louis, he started as soon as possible. The easiest way was from New York to Albany by boat, thence to Buffalo by canal, to Cleveland by steamer, to Portsmouth on the Ohio River, and then down the Ohio and up the Mississippi by steamboat. This trip took thirty days.

In St. Louis, which had then about 13,000 inhabitants. he soon got employment, but decided to go to New Orleans because of the approaching winter. He left St. Louis about Christmas, 1838, on foot, with his knapsack on his back; he crossed the Mississippi and walked along through the thinly settled forests of Illinois, the cane-brakes of Kentucky, and a part of Tennessee, where he fell in with two others going to the same destination. At the mouth of the Ohio they joined in buying a skiff and set out for New Orleans in it. They soon were caught by a steamer going their way and they boarded her and abandoned their skiff. Upon arriving in New Orleans the talk about Texas decided him to go farther west, and he arrived in Galveston in January, 1839. He stayed in Texas about a year and then returned to Illinois where he taught school for some time.

In the fall of 1841 he found an uninhabited island in the Missouri about three hundred miles above St. Louis, and he took up his solitary residence there. When the spring rise came it caused him to leave.

In 1844 he sailed for home, and while on this trip first learned that sets of dried plants might be sold. On his return to America and to St. Louis he began to collect and was aided by Dr. Engelmann in naming his specimens. He visited different parts of the country between Chicago and New Orleans for the purpose of collecting. Dr. E. gelmann commended him to Dr. Asa Gray, and he was furnished with the authority to accompany some troops which were being sent to Santa Fé, so that he had free transportation for himself and luggage. He returned to St Louis in the fall of 1847. In the spring of 1849 he started on another collecting trip to the West. He was unsuccessful, having lost most of his stock of drying papers in a flood, and he was forced to return to St. Louis. Upon his arrival here he found that all of his large collections and notes and journals had been destroyed in the great fire which burned much of the business section of the city during his absence. In 1849 he embarked for Panama, and after four months again returned to Arkansas, and finally went to Memphis, where he went into business. In 1854 he went to Venezuela and collected for four years, during this time exploring alone mountain ranges which were scarcely known at that time. He made very large collections, which are of great value. He returned to Missouri in 1864 and bought a tract of land in the town of Allenton, about thirty miles west of St. Louis. This he began to clear and cultivate in company with his half-brother, who was half-witted, and who always was dependent upon him. Here he remained for seven years, with the exception of a month spent in the Gray Herbarium, assisting in its arrangement. During this time Mr. Letterman became acquainted with him, and from 1870 to 1871 they met two or three times a week and nearly every Sunday with green plants to be identified. He seems to have collected but little in the vicinity, but was very familiar with the plants of the general neighborhood. After clearing his land and putting up his house, mostly with his own hands, he spent most of his time writing a book. This is undoubtedly his " Mechanism of the Universe," which was unfortunately published at his own expense later. Failing health forced him to dispose of his farm and remove to another climate. In 1871 he sold the farm and left for Europe, intending to live there the rest of his days. He, however, returned and settled at Wilmington, Delaware, in 1873. While here he finished his book and published it. Repeated attacks of rheumatism compelled him to seek a warmer climate, and he and his brother went to the island of Trinidad. They lived at Port of Spain, landing in June, 1877; here the remainder of his life was spent in making botanical observations and collecting, especially among the ferns. Advancing age restricted his

Fig. 14. House built by August Fendler in Allenton, Missouri, and occupied by him during his residence here from 1864 to 1871. The small ell has been added by subsequent owners.

efforts to the immediate neighborhood, and when this was exhausted he did but little. His death occurred in November, 1883.

An appreciation of his work from one who knew him best follows:

It is needless to say that Fendler was a quick and keen observer and an admirable collector. He had much literary taste, and had formed a very good literary style in English, as his descriptive letters show. He was excessively diffident and shy, but courteous and most amiable, gentle and delicately refined. Many species of his own discovery commemorate his name, as also a well-marked genus, *Fendlera*, a Saxifragaceous shrub which is winning its way into ornamental cultivation.[13]

[13] Gray, Asa, *Amer. Jour. Sci. and Arts*, 3d series, 29: 169, 1885.

Dr. F. Adolph Wislizenus came to America from Germany in 1835; he landed at New York and lived there for the next two years. In 1837 he went west, settling near Belleville, Ill. Two years later he came to St. Louis and lived in that city practically all the rest of his life. He is not known to have performed any botanical work in the vicinity of St. Louis, but he is included in the present paper because of having made a very considerable collection of plants in New Mexico, Mexico, and other parts of the great American arid plain. This collection was

FIG. 15. DR. ADOLPH WISLIZENUS; by permission of the St. Louis Academy of Sciences, from a photograph in their possession.

one of the first from the region visited, and is considered especially important because Dr. Wislizenus was one of the first to give an accurate. scientific account of the sections visited by him. This is especially true of Mexico, of which there were very erroneous and distorted ideas in the United States.

Dr. Frederick Adolphus Wislizenus[19] was born in 1810 at Koenigsee, in Schwarzburg-Rudolstadt, one of the numerous tiny German principalities of that period. He was the youngest of three children of a Protestant minister whose ancestors were said to have fled from Bohemia, victims of the religious fanaticism which resulted in the persecution of Hus and his followers.

[19] Engelmann, Geo. J., *Trans. St. Louis Acad. Sci.*, 5: 464–468, 1890.
Sargent, C. S., "Silva of North America," 6: 94, 1894.
Wislizenus, F. A., "Memoir of a Tour to Northern Mexico," 1–141, 1848.
POP. SCI. MONTHLY, 52: 643. 1898.

Wislizenus studied medicine at the University of Jena in 1828, and later at Göttingen and Würzburg. He was a member of the " Burschenschaft," but escaped arrest when that was broken up by the author-

He followed his friend and teacher, the great clinician Schoenlein, to Zürich and there joined an expedition to aid Mazzini in his struggle against Austrian rule ; but the Swiss troops disarmed them on the border so he was forced to return to his studies.

Wislizenus graduated in Zürich in 1834 and soon sailed for New York, where he began to practise his profession in 1835. Here he remained two years writing constantly for the German papers of the city. He then went west in 1837 and joined some of his fellow exiles who had settled in St. Clair County, Illinois. In 1839 he came to St. Louis and immediately seized an opportunity to accompany an expedition of the St. Louis Fur Company for trading with the Indians. He thus went far into the Northwestern country towards the source of the Green River in the Wind River Mountains. When the expedition started to return he joined a band of Flat-head and Nez Percé Indians. He thus crossed the Rocky Mountains to Utah and went as far as Fort Hall, the most southern post of the English trading company. Here he could find no guide to take him to California, so he returned: crossing the Green and the south fork of the Platte, he followed the Arkansas to Missouri. During this trip he had no facilities for making scientific observations and collections, so it was wholly without any such results.

On his return to St. Louis in 1840 he resumed his practise of medicine. He was identified with early efforts towards the establishment of an Academy of Science, and aided Dr. Engelmann in his efforts to found a botanic garden, and was an earnest worker in the Western Academy of Science. He soon gained a lucrative practice, but as soon as the opportunity offered he was again in the field. He joined a trading expedition to Mexico, well equipped this time with instruments and apparatus for scientific work. In Santa Fé they first learned of the war between Mexico and the United States, but Wislizenus obtained a pass and proceeded to Chihuahua, where he with other Americans was seized and imprisoned. He was sent to a small mountain town of the interior and there had ample opportunity to carry on his collecting and observations in the neighborhood during the winter. Upon the arrival of Col. Doniphan's troops in the spring he was released and accompanied them in a professional capacity until their disbanding at New Orleans in 1847, when he returned to St. Louis.

Senator Thomas H. Benton became interested in him and his experiences in Mexico, and finally was the cause of his being summoned to Washington and being requested to prepare for publication the results of his investigations. His resulting " Memoir of a Tour to Northern Mexico in 1846 and 1847 " was considered important enough so that the senate ordered 5,000 copies printed for distribution. This publication gave a good account of the country which was then much misunderstood and misrepresented, and resulted in correcting many erroneous ideas regarding that section of the American continent. It contained many very valuable data concerning the meteorology, geology, topography and botany of the region. Among the valuable results of this tour was a botanical collection containing many new plants which were classified and described by Dr. Geo. Engelmann, of St. Louis, who commemorated the valuable services of Wislizenus to science by applying his name to a new genus, *Wislizenia,* as well as to several of the new species of the collection.

Wislizenus again returned to St. Louis from Washington upon the completion of his report, and served faithfully during the cholera epidemic of 1849. As soon as this was over, however, he went to Constantinople in 1850 to bring back with him as his bride, Miss Lucy Crane, a sister-in-law of Hon. Geo. P. Marsh, whom he had met while in Washington. After visiting his old home in Thüringen and the large cities of the Old World, the two returned to the United States. Leaving his wife with her friends in the east, he went to Panama and California in search of a more desirable location. But he again returned to St. Louis and finally settled down permanently. He was one of the founders of the St. Louis Academy of Science and an active worker and one of the officers of the St. Louis Medical Society and of the Western Academy of Sciences. He was for many years president of the German Medical Society of St. Louis. His barometrical observations and his botanical and mineralogical collections, together with his memoir, are distinct additions to science. He was interested in meteorology from 1858 till his death, and in 1861 he commenced to study the atmospheric electricity with the belief that this would be of value in connection with meteorology. He discontinued this study, however, upon arriving at the conclusion that it was valueless in this connection—a fact which is now generally acknowledged. His last days were spent in seclusion, he being closely confined to the house by his infirmities and the loss of his sight. He died on September 22, 1889, in his eightieth year.

In 1851 there began a most important movement for the advancement of botany in St. Louis.[20] In that year, Mr. Henry Shaw, while on his last visit to Europe, first conceived the idea of establishing for himself a country estate on lines similar to those of many of the large English ones. In fact he had already started to build a home in the country district west of St. Louis.

This idea of a large private estate seems to have soon become changed to that of a botanical garden, for in 1857 he commenced active opera-

[20] Trelease, Wm., Mo. Bot. Garden Report, 1: 84–90, 1890. *Plant World,* 5: 1–4, 1902. "The Academy of Science of St. Louis," POP. SCI. MONTHLY, 62: 118–130, 1903. "The Missouri Botanical Garden," POP. SCI. MONTHLY, 62: 193–221, 1903.

tions to this end. He even at one time planned a grand school of botany with all the appendages and equipment necessary for a college of botany. This was modified in its first inception, but has been carried out to a degree. Very soon he built a botanical museum, bought herbaria and built greenhouses in which tender and exotic plants might be grown, while the grounds themselves were planted with many of the more hardy species. In 1859 he secured the passage of an act of the Missouri state legislature enabling him to deed or will to a board of trustees such property as he might wish, to be used for the maintenance of the Missouri Botanical Garden, as he prophetically named it. In 1885 he founded the Shaw School of Botany in connection with Washington University of St. Louis and provided for very close relations between the school and the garden. The estate deeded for the use of the garden was valued at about one million two hundred and fifty thousand dollars. This has increased very materially in value with the rapid rise in real estate in and about St. Louis. From the small beginnings of a private estate, the garden has developed until there were in cultivation in 1906, over seventeen thousand species and varieties of living plants; fifty-five thousand books and pamphlets in the library, including a very fine collection of pre-Linnæan works, and five hundred and sixty thousand sheets of dried specimens. The garden has issued eighteen annual reports, and is in exchange relations with nine hundred institutions interested in botany, gardening, horticulture or forestry. The library is one of the finest of the botanical libraries of the world, and all resources of the garden are placed at the free disposal of those capable of using them. Thus Mr. Shaw's life-work has reached its fruition, and a fitting memorial is rising steadily to more and more impressive proportions.

Henry Shaw[21] was born in Sheffield, England, July 24, 1800. He was the eldest of four children. His father was a manufacturer of grates, fire irons, etc., and owned a large establishment. Henry's early education was obtained at Thorne, a neighboring village, and his favorite place for study was an arbor in the garden. He was later transferred to Mill Hill, about twenty miles from London. This was termed a " dissenting " school, but was also considered one of the best private schools in the Kingdom. He remained here about six years, leaving probably in 1817, thus finishing his schooling. He studied while here considerable Greek, more Latin, more than the average amount of mathematics, French, and undoubtedly German, Italian and Spanish. With this scholastic training he began to assist his father at the home establishment for a year, after which he accompanied him to Canada. In this same year, 1818, his father sent him to New Orleans, mainly to investigate cotton raising. He stayed in Louisiana but a short time,

[21] Dimmock, Thos., *Mo. Bot. Garden Report*, 1: 7-25, 1890.

as he did not like the climate nor were there financial inducements for his doing so. He was now his own master and decided to go north and try his fortune in the then small and remote French trading post known as St. Louis. He accordingly embarked upon the *Maid of New Orleans,* and after a long and tedious voyage landed at St. Louis on May 3, 1819.

Fig. 16. Henry Shaw: from a watercolor painting at the Missouri Botanical Garden, by permission of the Director.

He began business on the second floor of a building which he found for rent, and for a time lived, cooked and sold his small stock of cutlery in this one room. The capital with which he bought his first stock of goods was furnished by his uncle. While Mr. Shaw's main object at this time was to make money, and while he denied himself many youthful enjoyments, he still did not thus deny himself beyond reasonable limits.

He had been succeeding in business, and when the balance sheet for 1839 was struck it showed to his own great surprise a net gain for the year of $25,000. His figures were gone over again and again until there could be no doubt of the fact. It seemed to him that " this was more money than any man in my circumstances ought to make in a single year." Accordingly, the following year, when opportunity offered, he closed out his business. At this time he was forty years of age, physically and mentally unimpaired, and vigorous, a free man, and the possessor of $250,000, equivalent to more than $1,000,000 at the present time.

In September 1840, Mr. Shaw made his first visit to Europe, stopping on his way at Rochester, New York, where his parents and sisters resided. He took an extended tour on the continent and, returning to St. Louis in the autumn of 1842, arranged his affairs for another absence in Europe. This lasted for about three years, during which time he visited all of the accessible European localities, together with Constantinople and Egypt. A journey to Palestine was prevented by the prevalence of the plague in that country.

Early in 1851 his last trip abroad was made, the first World's Fair being then held in London. While on this visit the idea first occurred to him to make a garden of his own, modeled after those which are so well known upon the great private estates of England. Mr. Shaw returned in December, 1851; the mansion at Tower Grove had been finished in 1849, and the one on the corner of Seventh and Locust streets was then being built. After this time he was in St. Louis, with the exception of short summer vacations at the Atlantic coast or the northern lakes. Seemingly a man of leisure, he was really a very busy man for the next thirty years, and was never an idler until compelled to be.

In 1857 the late Dr. Engelmann, who was then in Europe, was commissioned by Mr. Shaw to examine botanical gardens and to obtain such suggestions as he might think of value. About this time a correspondence was begun with Sir William J. Hooker, Director of Kew Gardens, who wrote on August 10, 1857:

> Very few appendages to a garden of this kind are of more importance for instruction than a library and economic museum, and these gradually increase like a rolling snowball.

Accordingly, Mr. Shaw in 1858–9 erected a building for this purpose. The selection of books was entrusted largely to Dr. Engelmann in consultation with Hooker, Decaisne, Alexander Braun and others of his botanical friends. At the same time Dr. Engelmann urged upon Mr. Shaw the purchase of the herbarium of the recently deceased Professor Bernhardi, of Erfurth, Germany, which was offered at a very small price. Hooker wrote January 1, 1858:

> He [Engelmann] tells me of the herbarium of the late Dr. Bernhardi, of Erfurth, which he expects to buy for St. Louis. That ought to be a good commencement for the more scientific part of the establishment. . . . The state ought to feel that it owes you much for so much public spirit, and so well directed.

Mr. Shaw has told that he at one time planned a grand school of botany, with residences for the faculty, laboratories, etc., opposite the main gate; but he abandoned the project because of the advice of Dr. Asa Gray.

In 1866 Mr. Shaw secured the services of Mr. James Gurney from the Royal Botanical Gardens of London, whose practical experience and faithfulness contributed very largely to make the Garden and Tower Grove Park what they are to-day. Mr. Shaw, however, never abandoned his personal supervision, and he thus spent the last twenty-five years of his life perfecting what he had begun. Until the summer of 1885 he had not been out of St. Louis, except to drive out to dine with a friend, for about twenty years. At this time the hot weather caused a failure of his usual good health, and he went to northern Illinois and Wisconsin for some time. He returned much improved and resumed his accustomed avocations with renewed vigor.

On the twenty-fourth of July, 1889, he received numerous visitors who congratulated him upon the beginning of his ninetieth year. Although weak, he was able to meet them in the drawing-room, and his mind was as clear as ever. This, however, was his last public appearance. An attack of malaria resulted in his death on August 25. On

FIG. 17. MR. GEO. W. LETTERMAN.

Saturday, August 31, he was laid to rest in the mausoleum which had been already prepared in the midst of the garden which he had created —not only for himself, but for all succeeding generations.

Mr. G. W. Letterman is one of the few persons who have worked upon botany in the vicinity of St. Louis during their whole lifetime. Mr. Letterman has worked especially in Missouri, but is also very familiar with the plants of the region included in eastern and northern Texas, Louisiana, Arkansas and Indian Territory. He has accumulated a very large herbarium, in which the flora of St. Louis is represented probably better than in any other private herbarium.

George Washington Letterman,[22] the son of John and Charlotte (Blair) Letterman, was born near Bellefonte, Center County, Pennsylvania, of a family which had lived for three generations in Pennsylvania, his father being of Dutch, and his mother of Irish descent. From the public school he entered the State College in Center County, but left before graduation to join the Union Army, in which he enlisted as a private; serving until the end of the war he was mustered out of the service with the rank of captain of volunteers. After crossing the plains to New Mexico in 1866, he returned to Pennsylvania, and then going west again to Kansas, with the idea of becoming a farmer in that state, he finally, in 1869, settled in Allenton, Missouri, a railroad hamlet about

[22] Sargent, C. S., "Silva of North America," 13: 79–80, 1902.

thirty miles west of St. Louis. Here Mr. Letterman taught in the public schools uninterruptedly for twenty years, and then for two years served as superintendent of schools in St. Louis County. Shortly after settling in Allenton Mr. Letterman met August Fendler, the botanist, who had a farm at this time in the neighborhood. This meeting with Fendler stimulated his interest in plants, especially in trees, and led to an acquaintance with Dr. Engelmann, for whom Letterman made large collections of plants in the neighborhood of Allenton, with many notes on the oaks and hickories. In 1880 he was appointed a special agent of the Census Department of the United States, to collect information about the trees and forests of Missouri, Arkansas, western Louisiana and eastern Texas, and later he was employed as an agent of the American Museum of Natural History in New York, to collect specimens of the trees of the same region for the Jesup collection of North American woods. The distribution of the trees of this region before Mr. Letterman's travels was little known, and much useful information concerning them was first gathered by him. Of his numerous discoveries species of *Vernonia*, *Poa* and *Stipa* commemorate the name of Letterman.

The above account is taken verbatim from Sargent's " Silva of North America," as it is the only authentic account of Mr. Letterman's life available. Mr. Letterman still lives at Allenton, Missouri, and is carrying on his botanical work. From the accounts of those in a position to know, his herbarium is very large, and at the present time probably contains as complete a representation of the St. Louis flora as any other, with the possible exception of the Eggert collection, which, however, can hardly surpass it. Mr. Letterman is connected with the local botanical societies, and is well known by the botanical workers of the city.

One man who has left an enduring impression upon botany, although his life work was along other lines, was Dr. Charles Valentine Riley.[23] Dr. Riley was born at Chelsea, London, September 18, 1843. His boyhood was spent at Walton-on-Thames, where he became acquainted with W. C. Hewitson, the author of a work on butterflies. This acquaintance undoubtedly turned his inclinations towards entomology. He studied for three years in the school at Dieppe and afterwards at Bonn. His teacher at the latter place urged him to study art at Paris, but this was not done. At the age of seventeen he emigrated to Illinois and when about twenty-one went to Chicago as reporter and editor for the *Prairie Farmer*. He was for six months in an Illinois regiment during the latter part of the Rebellion. He attained such success as an entomologist that he was made State Entomologist for Missouri in 1868, and he held this office until 1877, when he went to Washington in the government service. During this period he and his assistants, Miss Mary E. Murtfeldt and Mr. Otto Lugger, worked out two cases of the relation of insects to plants which are of more than ordinary interest.

In 1863 there were first noted in France the ravages of the Ameri-

[23] Howard, L. O., *Proc. Soc. Prom. Agric. Sci.*, 17: 108–112, 1896.

can *Phylloxera* upon the tender European varieties of grapes. These injuries became so serious that in 1872 the trouble was known not only in France, but in Portugal, Switzerland, Germany and England, and the entire grape and wine industry of Europe was threatened with annihilation. Riley became much interested in the problem of controlling the pest and finally hit upon the plan of grafting the susceptible European varieties upon roots of the resistant American species. This simple expedient undoubtedly saved the grape industry of Europe and also incidentally prevented a tremendous loss of money.

The second case was one of purely scientific value and interest. Dr. George Engelmann had noted that the character of the pollen of *Yucca* indicated that pollination of the flowers must be accomplished by some kind of an insect. Riley took up this hint and finally, with the aid of his assistants, discovered that the pollination wa actually performed by the *Pronuba* and *Prodoxus* moths. This line of work was continued for twenty years, and a series of publications upon it issued at various times during this period.

Incidentally his work was of interest to botanists in many other cases, but these two seem especially noteworthy. He won an enviable reputation among entomologists the world over. He died the latter part of the year 1895.

Because of her botanical work, as well as her association with Dr. Riley in working out the pollination of *Yucca* and other problems, Miss Mary E. Murtfeldt deserves mention. In 1885 Professor S. M. Tracy, then of Columbia, Missouri, published a list[24] of the plants of the state. In this list one finds many species from the vicinity of St. Louis credited to " Murtfeldt " as their collector. These specimens were collected by Miss Murtfeldt not long before the publication of the " Tracy " list and are still in her possession, forming a collection of about 500 numbers. Miss Murtfeldt's first scientific work was in botanical lines, but this later changed to entomology, her botanical knowledge being indispensable in following out the life histories of new or little known insects upon their host plants. Many of her later botanical specimens are of much interest from the entomological standpoint and were prepared for that purpose alone. Miss Murtfeldt is well known among entomologists for her work, which has been mostly of this nature.

In 1874 Mr. Henry Eggert, as he was known, came to St. Louis, went into business, and began the study of the local flora and the formation of an herbarium which probably represented the flora of that vicinity at the time of his death, the best of any in existence. Eggert came to America from Prussia when about thirty years of age; he had already collected and studied the plants of different sections in Europe,

[21] Tracy, S. M., " Flora of Missouri," *Mo. State Hort. Soc. Report* (Appendix), 1–106, 1885.

and his work about St. Louis seems to have been simply a continuation along similar lines to that already done in Europe. Although he lived in St. Louis and in later years in East St. Louis, he seems to have been somewhat of a hermit, and was not understood, or even comparatively well known, by his neighbors. He seems to have been an enthusiast upon botany, and his botanical collection was apparently his one luxury and hobby.

Heinrich Karl Daniel Eggert[25] was born March 3, 1841, in the town of Osterwieck, Prussia. He was educated at a seminary in Halberstadt, and became a teacher in the public schools of the neighboring

FIG. 18.   THE EGGERT HOUSE IN EAST ST. LOUIS, ILLINOIS; practically as it was at the time of the death of Henry Eggert.

city of Magdeburg. He early became interested in the study of plants, and before leaving Europe he had made botanical collections in the Harz Mountains and on short journeys to Kreuznach and in Bohemia. Dissatisfied with the small salary of a German school teacher, Eggert came to America in 1873, and for a few months worked on a farm in southern New York. From New York he went to St. Louis, where he remained for a number of years and then removed across the river to East St. Louis, where he lived the rest of his lifetime.

The first work which he seems to have taken up in St. Louis was that of carrying papers for the local press. He carried papers for about twenty years, handling both a morning and an evening one. He worked early and late, never sparing himself and always living by himself in a secluded manner. Comparatively few persons ever saw the in-

[25] Sargent, C. S., "Silva of North America," 13: 51-52, 1902.

terior of his house, and still fewer were on really friendly terms with him, as we ordinarily use that phrase. While he had but little to do with his neighbors he never seems to have had any enemies.

Eggert's first start in making more money than usual was at the time of the great outbreak of the American *Phylloxera* in the vineyards of Europe, destroying immense numbers of the vines and threatening the entire wine and grape industry of Europe. It was finally discovered that the American native grapes might be used as stocks upon which to graft the more susceptible European varieties, so that a vine was obtained which had roots of the American resistant species with the top of some desirable but susceptible European species. This work resulted in an immense demand for the seed of some of our native species of grapes. Eggert's knowledge of botany led to his being recommended as a suitable person from whom to get these seeds. For at least two or three years he made a business of collecting and selling them to foreign countries. The business was quite remunerative and in the proper season he is said to have made several hundred dollars a month in this way. He seems to have kept up his carrying of papers at the same time. At first he carried them on his back, taking immense loads in a bag slung over his shoulder. As his business grew he bought a horse and wagon and still later he employed others, so that at one time he conducted a considerable business of this kind. He never relinquished his botanical work, and in early days he collected specimens for sale to botanists and for use in colleges and schools, thus making some little money. In later years his left arm and hand became affected with a partial paralysis which he attributed to his severe work in carrying such heavy weights of papers slung over that shoulder.

His money he invested in farms and similar property, and he succeeded in amassing considerable property. In his personal habits he was always very frugal, his only luxury seeming to have been his botanical collecting. In 1896 he sent to Germany for his nephew, August Eggert, and turned his greenhouses over to him to run. This nephew lived more or less intimately with him. Mr. Eggert was always of a peculiar disposition, apparently being constantly in fear of some attempt upon his life. He had hallucinations in which he thought every one had designs upon his life, and these became worse as he grew older. His mind was undoubtedly unbalanced, and on the night of April 18, 1904, he shot himself with a revolver.

As mentioned above, Eggert early learned botany and collected extensively all of his life. He collected assiduously all around St. Louis for a considerable distance, and his collection probably represented the flora of this district better and more completely than any other ever made. He also went on collecting trips to various parts of Missouri, Illinois, Arkansas, Alabama, Tennessee and Texas, and the southeastern states. He seemed to possess a genuine love for botany, and his determinations seem to have been, as a rule, correct beyond the ordinary. He was a charter member of the Engelmann Botanical Club, and was its first vice-president. He was also a member of the International Association of Botanists, and was made one of its vice-presidents.

Personallly, he seems to have had no enemies; he always remembered an injury, either real or fancied, and was unstinting in his expression of dislike for those who had in any way incurred his displeasure. His love of botany and his fine herbarium made him well known to the local botanists, yet he never seems to have been on really intimate terms with many of them. He was always ready to exchange specimens of rare plants or local species, and his herbarium was thus greatly enlarged by exchange from other countries as well as from all parts of the United States. During early days he collected specimens for the purpose of selling them, but as he grew older he could rarely be induced to sell his specimeñs, preferring to exchange.

His herbarium at his death was estimated to contain about 60,000 specimens, and was considered very valuable. It was acquired by the Missouri Botanical Garden, and is at present being incorporated with the herbarium of that institution as rapidly as possible. His herbarium is especially valuable for the reason that it was the basis of a local flora published by Eggert in 1891 under the title " Catalogue of the Phænogamous and Vascular Cryptogamous Plants of the Vicinity of St. Louis, Mo." His preface is characteristic and self-explanatory, so that it may well be given:

Since[20] the publication of Mr. Geyer's catalogue of the Plants of Illinois and Missouri, about 1842, no other effort has been made to publish a list of plants growing in the vicinity of St. Louis but my own partial lists of species found in former years. I hope my present catalogue of Plants growing in a radius of about 40 miles around St. Louis will be welcome to botanists until a local flora is published.

Since 1874 I have systematically looked over the ground in all directions, so that very few plants will have escaped my observation; but as I could only go out one day at a time, in places too far off from railroads, there still may be found something new. Railroads also will bring new immigrants from other regions when some of our own plants may have vanished, so that it will be a very important matter for later botanists to know what in former years was growing here. This idea mostly led me to have this catalogue printed.

With the exception of a few plants reported to me by Mr. Letterman, of Allenton, Mo., all plants are collected by myself. The catalogue contains nearly 1,100 different species and varieties, so that St. Louis need not be ashamed of her flora.

This catalogue of Mr. Eggert's is by far the best and most nearly complete list of our plants which has yet appeared. Besides the above mentioned catalogue, a number of small lists of desiderata were dis-

[20] Eggert, Henry, "Catalogue of the Phænogamous and Vascular Cryptogamous Plants in the Vicinity of St. Louis, Mo.," 1–16, 1891.

tributed to Eggert's correspondents for a number of years. Aside from these he published absolutely nothing, so far as now known. Exact localities were not given either in his lists or upon the labels accompanying his specimens, but he is known to have kept a note-book in which all such data were given. This note-book disappeared during the changes following his death, and thus much valuable and intimate knowledge of our flora was lost. As mentioned above, his entire herbarium is now in the possession of the Missouri Botanical Garden, where it will receive the best of care and will be accessible to all botanists desiring to use it.

One of the more recent collectors who have worked in and about St. Louis, especially upon the fleshy fungi, is Dr. N. M. Glatfelter.

Dr. Noah M. Glatfelter was born in York County, Pennsylvania, on November 28, 1837. He lived on a farm until he was seventeen years of age, when he began teaching school. He finished seven terms, and during the time attended successively the York County Academy, Lancaster County Normal School, and Franklin and Marshall College at Lancaster, Pa., for two thirds of the sophomore year. He then commenced the study of medicine with Dr. John L. Atlee, of Lancaster. In 1862 he attended the medical lectures at the University of Pennsylvania, and graduated from the same institution in 1864. He then received a commission from President Lincoln as Assistant Surgeon of United States Volunteers. In 1867 he left the army in Dakota territory. Ever since that time he has practised medicine in and near St. Louis.

FIG. 19.   DR. N. M. GLATFELTER; about 1900.

About 1889 Dr. Glatfelter commenced collecting the herbaceous plants in the vicinity of St. Louis and obtained specimens of most of the species of the district. This herbarium is still in the collector's possession. From 1892 to 1898 he gave special attention to the willows of St. Louis, and contributed papers on the venation of *Salix*, on *Salix* hybrids, on *Salix longipes* and on the relations between *Salix nigra* and *S. amygdaloides*.

In 1898 he became interested in the collection and study of the Hymenomycetes. This has led to the accumulation of about five hundred species, making quite an exhaustive collection of these fungi. This work is being continued and has already resulted in the discovery of a number of species new to science, several of which have been named in honor of their discoverer. This material has been submitted to Professor Chas. H. Peck, so it is authoritatively named.

In 1906 a list of this collection was published by the St. Louis Academy of Science.[27] The specimens are mostly in Dr. Glatfelter's private herbarium. Collecting has also been done in Pennsylvania in 1899, 1905 and 1906, and somewhat in other states. The herbaceous herbarium has been increased by exchanges, so that it numbers over 4,000 species. Dr. Glatfelter is a member of the local botanical societies and is still collecting the fleshy fungi, to which he is giving most of his attention.

The more recent botanical workers of St. Louis we find grouped into two distinct bodies; the staff of the Shaw School of Botany, and of the Missouri Botanical Garden, and the investigators of the Mississippi Valley Laboratory of the United States Department of Agriculture. In the former group, which has existed for the longer time, the following persons should be mentioned: Dr. William Trelease, director of the Missouri Botanical Garden since the death of Mr. Shaw, and also professor of botany in the Shaw School of Botany. Besides administering the affairs of these two institutions, and bringing them to their present development and efficiency, he has published many scientific papers; the earliest ones were concerned with fungi and various plant diseases; then the pollination of flowers was taken up; and of late years his work has been in the systematic revision of certain groups, such as the genera *Acer, Rumex, Yucca,* etc. Under his management the botanical garden has issued eighteen annual reports of scientific material, which have given that institution a name for scientific research, although it can hardly even yet be said to have fairly emerged from the preparatory stage of its development. Associated very closely with Doctor Trelease since 1894 is Mr. H. C. Irish, who has had general charge of the grounds, greenhouses and outdoor planting. Mr. Irish has published papers on horticultural subjects, including a scientific revision of the genus *Capsicum,* and of the " garden bean," and has in preparation another extensive paper along similar lines. Mr. C. H. Thompson has been connected with the garden for a number of years, and is engaged also upon scientific investigations. Dr. J. A. Harris, librarian of the garden, has published a number of scientific papers, and is engaged upon others, in the preparation of which the extensive and excellent library facilities of the garden are being fully employed. Others who have been connected with the garden staff, and who are now well known scientifically, are Dr. L. H. Pammel, Dr. H. J. Webber and

[27] Glatfelter, N. M., " Preliminary List of Higher Fungi Collected in the Vicinity of St. Louis, Mo., from 1898 to 1905," *Trans. Acad. Sci. St. Louis,* 16: 33–94, 1906.

J. B. S. Norton, all of whom worked more or less upon the fungi of the locality while at the garden. Dr. S. M. Coulter, assistant professor of botany in the Shaw School of Botany, has, ever since coming to St. Louis, been working upon ecological problems.

The second group of botanists is a small one, of whom the following have been more or less intimately connected with the local work being carried upon the flora of the vicinity: Dr. Hermann von Schrenk, in charge of the Mississippi Valley Laboratory until its removal to Washington in 1907, has published a number of scientific papers dealing with the diseases of forest trees and of timber. Some of these were worked out from material collected around St. Louis, either partially or entirely. Dr. von Schrenk continues his work at St. Louis, having severed his relations with the United States Department of Agriculture upon the removal of the Mississippi Valley Laboratory from St. Louis to Washington. Drs. G. G. Hedgcock and Perley Spaulding, assistants of Dr. von Schrenk, were also engaged upon prob'ms relating to the diseases of fruit and forest trees. All three have collected the fungi of the vicinity, and have been intimately connected with the botanical activities of the place.

Besides the above workers should be mentioned Mr. John Kellogg, long employed by the garden, who is very familiar with the local flora, and has a very good private herbarium; Dr. N. L. T. Nelson, who is collecting the mosses of the vicinity; Mr. H. M. T. Hus, who is collecting the algæ; and numbers of others who have collected in the locality at various times.

BIOLOGISTS AND THEIR WORLD

An Arno Press Collection

Adler, Kraig, editor. **Early Herpetological Studies and Surveys in the Eastern United States.** 1978

Adler, Kraig, editor. **Herpetological Explorations of the Great American West.** Two vols. 1978

Agassiz, [Jean] Louis [Rodolphe]. **Contributions to the Natural History of the United States of America.** Four vols. in two. 1857/1860/1862

Allard, Dean Conrad, Jr. **Spencer Fullerton Baird and the U.S. Fish Commission.** 1978

Altum, Bernard. **Der Vogel Und Sein Leben.** 1868

Azara, Don Felix de. **Apuntamientos Para La Historia Natural De Los Quadrúpedos Del Paragüay Y Rio De La Plata.** Two vols. in one. 1802

Baer, Karl Ernst v[on]. **Reden Gehalten in Wissenschaftlichen Versammlungen und Kleinere Aufsaetze Vermischten Inhalts.** Three vols. in two. 1864/1876/1873

Barrett-Hamilton, Gerald E. H. and Martin A. C. Hinton. **A History of British Mammals.** Two vols. 1910-1921

Boulenger, G[eorge] A[lbert]. **The Tailless Batrachians of Europe.** Two parts in one. 1897/1898

Brocchi, [Paul]. **Mission Scientifique Au Mexique Et Dans L'Amérique Centrale,...** Étude Des Batraciens De L'Amérique Centrale. 1882

Buffon, [Georges L. L.]. **The History of Singing Birds.** Translated from the French. 1791

Buffon, [Georges L. L.]. **The Natural History of Oviparous Quadrupeds and Serpents.** Translated by Robert Kerr. Four vols. in two. 1802

The Cabinet of Natural History and American Rural Sports. Three vols. in one. 1830/1832/1833

Candolle, A[ugustin] P[yramus] de and K[urt] Sprengel. **Elements of the Philosophy of Plants.** Translated from the German. 1821

Cassin, John. **United States Exploring Expedition During the Years 1838, 1839, 1840, 1841, 1842 Under the Command of Charles Wilkes, U. S. N.:** Mammalogy and Ornithology. Two vols. 1858

Chapman, Frank M. **Essays in South American Ornithogeography.** Edited by Keir B. Sterling. 1978

Cope, Edward D[rinker]. **The Vertebrata of the Tertiary Formations of the West.** One vol. in two. 1883

Cuvier, [Georges]. **The Class Mammalia.** Five vols. 1827

Donovan, E[dward]. **The Natural History of British Fishes.** Five vols. in two. 1802/1803/1804/1806/1808

Duméril, Auguste [Henri André], [Marie-Firmin] Bocourt and [François] Mocquard. **Mission Scientifique Au Mexique Et Dans L'Amérique Centrale,...** Étude Sur Les Reptiles. Two vols. 1870-1909

Flower, William Henry and Richard Lydekker. **An Introduction to the Study of Mammals, Living and Extinct.** 1891

Forbush, Edward Howe. **Birds of Massachusetts and Other New England States.** Three vols. 1925/1927/1929

Girard, Charles. **United States Exploring Expedition During the Years 1838, 1839, 1840, 1841, 1842 Under the Command of Charles Wilkes, U. S. N.: Herpetology.** Two vols. 1858

Grinnell, Joseph. An Account of the Mammals and Birds of the Lower Colorado Valley. 1914

Howard, H[enry] Eliot. **Territory in Bird Life.** 1920

Hume, Edgar Erskine. **Ornithologists of the United States Army Medical Corps.** 1942

Huxley, Leonard. **Life and Letters of Sir Joseph Dalton Hooker.** Two vols. 1918

LeConte, John L. and George H. Horn. **Classification of the Coleoptera of North America.** 1883

Linsley, E. Gorton, editor. **Beetles From the Early Russian Explorations of the West Coast of North America, 1815-1857.** 1978

Linsley, E. Gorton, editor. **The Principal Contributions of Henry Walter Bates to a Knowledge of the Butterflies and Longicorn Beetles of the Amazon Valley.** 1978

Murray, Andrew. **The Geographical Distribution of Mammals.** 1866

Osborn, Henry Fairfield. **Cope:** Master Naturalist. 1931

Peale, Titian R. **United States Exploring Expedition During the Years 1838, 1839, 1840, 1841, 1842 Under the Command of Charles Wilkes, U. S. N.:** Mammalogy and Ornithology. 1848

Phillips, John C. **American Game Mammals and Birds.** 1930

Rafinesque: Autobiography and Lives. 1978

Ray, John. **Synopsis Methodica Animalium Quadrupedum et Serpentini Generis.** 1693

Ray, John. **Synopsis Methodica Avium & Piscium.** [Edited by William Derham]. 1713

Richardson, John. **Fauna Boreali-Americana;** Or the Zoology of the Northern Parts of British America, Part Third: The Fish. 1836

Richardson, John, William Swainson and William Kirby. **Fauna Boreali-Americana;** Or the Zoology of the Northern Parts of British America, Part Four: Insecta, 1837

Riley, Charles V[alentine]. **[Nine] Annual Report[s] on the Noxious, Beneficial and Other Insects of the State of Missouri, 1869-1877** *and* **General Index and Supplement.** Ten vols. in three. 1869-1877/1881

Say, Thomas. **The Complete Writings of Thomas Say on the Entomology of North America.** Edited by John L. LeConte. Two vols. 1859

Schuchert, Charles and Clara Mae LeVene. **O. C. Marsh:** Pioneer in Paleontology. 1940

Sclater, William Lutley and Philip Lutley Sclater. **The Geography of Mammals.** 1899

Seton, Ernest Thompson. **Trail of an Artist-Naturalist:** The Autobiography of Ernest Thompson Seton. 1940

Smith, James Edward. **A Selection of the Correspondence of Linnaeus and Other Naturalists, From the Original Manuscripts.** Two vols. 1821

Stuckey, Ronald L., editor. **Development of Botany in Selected Regions of North America Before 1900.** 1978

Stuckey, Ronald L., editor. **Essays on North American Plant Geography From the Nineteenth Century.** 1978

Stuckey, Ronald L., editor. **Scientific Publications of Charles Wilkins Short.** 1978

Swammerdam, John. **The Book of Nature.** 1758

Wadland, John Henry. **Ernest Thompson Seton:** Man in Nature and the Progressive Era, 1880-1915. 1978

Waterhouse, G[eorge] R[obert]. **A Natural History of the Mammalia.** Two vols. 1846/1848

Weiss, Harry B. and Grace M. Ziegler. **Thomas Say:** Early American Naturalist. 1931

Wheeler, Geo[rge] M., [editor]. **[Reports Upon Insects Collected During Geographical and Geological Explorations and Surveys West of the One Hundredth Meridian During the Years 1872, 1873, and 1874].** 1875

Willughby, Francis. **De Historia Piscium** *and* **Icthyographia Ad Amplisimum Virum Dnum., Samuelem Pepys, Presidem Soc. Reg.** Two vols. in one. 1686/1685

Youmans, William Jay, editor. **Pioneers of Science in America.** Revised edition. 1896